Building Knowledge Regions in North America

Building Knowledge Regions in North America

Emerging Technology Innovation Poles

Leonel Corona

National University of Mexico

Jérôme Doutriaux

University of Ottawa, Canada

Sarfraz A. Mian

State University of New York, Oswego, USA

Edward Elgar

Cheltenham, UK • Northampton, MA, USA

Published by
Edward Elgar Publishing Limited
Glensanda House
Montpellier Parade
Cheltenham
Glos GL50 1UA
UK

Edward Elgar Publishing, Inc.
136 West Street
Suite 202
Northampton
Massachusetts 01060
USA

A catalogue record for this book
is available from the British Library

Library of Congress Cataloguing in Publication Data

Corona Treviño, Leonel.
Building knowledge regions in North America : emerging technology innovation poles / Leonel Corona, Jérôme Doutriaux, Sarfraz A. Mian.
p. cm.
Includes bibliographical references.
1. Technological innovations – Economic aspects – North America. 2. Technology and state – North America. 3. Research parks – North America. I. Doutriaux, Jérôme, 1945– II. Mian, Sarfraz, A. 1952– III. Title.

HC95.Z9T4+
338.97'06 – dc22

ISBN–13: 978 1 84542 430 5
ISBN–10: 1 84542 430 1

Printed and bound in Great Britain by MPG Books Ltd, Bodmin, Cornwall

Contents

Figures

Tables

Acronyms and abbreviations

$bn	Billions of dollars (one thousand million dollars)
$	United States dollar
Angel, angel investor	A wealthy individual, often a very successful high-tech entrepreneur, providing venture capital funds to a start-up or an entrepreneur
CA	Canada
C$	Canadian dollar
GATT	General Agreement on Trade and Tariffs
GDP	Gross Domestic Product
GSP	Gross State Product
GNP	Gross National Product
GERD	Gross Expenditures on Research and Development
IBM	International Business Machines Corporation
ICT	Information and Communications Technologies
IPO	Initial Public Offering
IT	Information Technologies
MNF	Multinational firm
Monarca (Monarch)	Research project symbol for the NAFTA countries (butterfly that flies all over North America)
NAFTA	North American Free Trade Agreement
NIS	National Innovation System
NSF	National Science Foundation (USA)
OECD	Organisation for Economic Cooperation and Development
PPP$	Purchasing Power Parity Dollars
RC	Research Center
RC&U	Research Centers and Universities
R&D	Research and Development
RIS	Regional Innovation System
SME	Small and Medium Enterprise
S&E	Scientists and engineers
TBED	Technology-based Economic Development
TBF	Technology-based firm
TIP	Technology innovation pole
UK	United Kingdom
US	United States of America

VC fund	Venture capital fund
VC	Venture capitalist
WTO	World Trade Organization

ACRONYMS AND ABBREVIATIONS SPECIFIC TO UNITED STATES

AARC	Alternative Agricultural Research and Commercialization
AITP	Aerospace Industry Technology Program
ARCH	Argonne National Lab, Chicago
ARPA	Advanced Research Project Agency
ATP	Advanced Technology Programs
AURP	Association of University Research Parks
BIRL	Basic Industrial Research Lab
BTC	Business/Technology Center
CIT	Center for Innovation Technologies
CRADA	Cooperative Research and Development Act
EDA	Economic Development Assistance
EPSCoR	Experimental Program to Stimulate Competitive Research
ERC	Engineering Research Center
ES	Extension Service
ETI	Environmental Technology Initiative
FLC	Federal Laboratory Consortium
FTE	Full Time Equivalent
FTTA	Federal Technology Transfer Act
GE	General Electric
GNI	Gross National Income
HRAP	Human Resources Assessment Project
IVHS	Intelligent Vehicle Highway System
MANTECH	Manufacturing Technology Program
MEP	Manufacturing Extension Partnership
MG&E	Madison Gas and Electric
MR	Magnetic Resonance
MTC	Manufacturing Technology Centers
NAICS	North American Industry Classification System
NASA	National Aeronautics and Space Administration
NCHCP	National Consortium for High Performance Computing
NCRA	National Cooperative Research Act
NGA	National Governors Association
NIH	National Institute of Health
NIST	National Institute of Standards and Technology

NRV	New River Valley
NSB	National Science Board
NSF	National Science Foundation
NSTC	National Science and Technology Council
NTTC	National Technology Transfer Center
NU	Northwestern University
NUERP	Northwestern University Evanston Research Park
NWAC	National Workforce Assistance Collaborative
ONR	Office of Naval Research
OSTP	Office of Science and Technology Policy
PMSA	Primary Metropolitan Statistical Area
RenTEC	Rensselaer Technology Development Council
RPI	Rensselaer Polytechnic Institute
RTP	Rensselaer Technology Park
RTTC	Regional Technology Transfer Center
SBDC	Small Business Development Center
SBI	Small Business Initiative
SBIR	Small Business Innovation Research
SLIPTI	State & Local Initiatives for Productivity, Technology and Innovation
SSET	State Science, Engineering and Technology
STC	Science and Technology Center
STEP	State Technology Extension Program
STS	State Technical Services
STTR	Small Business Technology Transfer Program
SUNY	State University of New York
TDOs	Technology Development Organizations
TIP	Technology innovation pole
TRP	Technology Reinvestment Project
TTYs	Text Telecommunications Devices
TUO	Technology Utilization Office
UIR	University Industry Relations
URP	University Research Park
USIP	United States Innovation Partnership
UW	University of Wisconsin
VT	Virginia Tech, Virginia Polytechnic Institute
VTCRC	Virginia Tech – Corporate Research Center
WARF	Wisconsin Alumni Research Foundation

ACRONYMS AND ABBREVIATIONS SPECIFIC TO CANADA

AT&T	American Telephone and Telegraph
BC	British Columbia
BCIT	British Columbia Institute of Technology
BNR	Bell Northern Research
CANARIE	Canadian Advanced Network for Research, Industry and Education
CEIM	Centre d'entreprises et d'innovation de Montréal
CIHR	Canadian Institute for Health Research
CISTI	Canadian Institute for Scientific and Technical Information
CLD	Centres Locaux de Développement
CMA	Census Metropolitan Area
Crown Corporation	A corporation owned by the government, organized and run as if it were a private company
CR&DA	Calgary Research and Development Authority
CRC	Communication Research Centre
CRC-IC	Communication Research Centre Innovation Centre
CREDEQ	Centre Régional de Développement d'Entreprises de Québec
CRIQ	Centre de recherche industrielle, Québec
CTI	Calgary Technologies Inc.
CTN	Canadian Technology Network
FTE	Full time equivalent
FY	Fiscal year
GATIQ	Groupe d'action pour l'avancement technologique et industriel de la région de québec
IBM	International Business Machines Corporation
ICT	Information and Communications Technology
INO	Institut National d'Optique, Québec
INRS	Institut National de la Recherche Scientifique
IPF	Industry Partnership Facility (NRC)
IRAP	Industrial Research Assistance Programme
IT	Information Technology
ITA	Industry Technology Advisors (NRC)
JDS	Predecessor of JDS-Uniphase
MRC	Medical Research Council
NAIT	Northern Alberta Institute of Technology
NCE	Network of Centres of Excellence
NRC	National Research Council, www.nrc.gc.ca
NRC-IPF	National Research Council Industry Partnership Facility
NSERC	Natural Science and Engineering Research Council

OBIC	Ottawa Biotechnology Incubation Centre
OCRI	Ottawa Centre for Research and Innovation (formerly Ottawa Carleton Research Institute), www.ocri.ca
OED	Ottawa Economic Development
OLSC	Ottawa Life Sciences Council
OLSTP	Ottawa Life Sciences Technology Park
RIM	Research In Motion
SDQM	Société Québécoise de la Main d'Oeuvre
SFU	Simon Fraser University
SOCO	Saskatchewan Opportunities Corporation
SSHRC	Social Science and Humanities Research Council
TEC	Technology Enterprise Centre, University of Calgary
TRLabs	Telecommunication Research Laboratories
UBC	University of British Columbia
UQAM	Université du Québec à Montréal
UST	University of Saskatchewan Technologies Inc.

ACRONYMS AND ABBREVIATIONS SPECIFIC TO MEXICO

AI	Academia de Ingeniería (Academy of Engineering)
AMIEPAT	Asociación Mexicana de incubadora de Empresas y Parques Tecnológicos (Mexican Association of Technology Parks and Enterprise Incubators)
AMIRE	Asociación Mexicana de Incubadoras y Redes Empresariales (Mexican Association of Management Incubation)
ANUIES	Asociación Nacional de Universidades e Instituciones de Educación Superior (National Association of Mexican Universities)
AVANCE	Alto Valor Agregado en Negocios (Conacyt's high valued added business program)
CCF	Centro de Ciencias Físicas de la UNAM (Physical Science Centre)
CECIMAC	Centro de Investigación de la Materia Condensada de la UNAM (Condensate Materials Research Centre)
CEMIT	Centro de Empresas para la Innovación Tecnológica (Center for Businesses of Technological Innovation)
CENAM	Centro Nacional de Meteorología (The National Standards Centre)
CENIT	Centro de Negocios e Incubación Tecnológica (Business Centre and Technology Incubation)

CETAF	Centro Empresarial Tecnológico Agropecuario y Forestal (Forestry and Agribusiness Technology Center)
CGCyT	Consejo General de Ciencia y Tecnología (General Science and Technology Council)
CIATEC	Centro de Investigación y Asesoría Tecnológica en Cuero y Calzado (Center for Research and Technological Consultancy in leather and footwear)
CIBNOR	Centro de Investigaciones Biológicas del Noroeste (Biology Research Centre of the North East)
CIATEQ	Centro de Investigación y Tecnología Avanzada del Estado de Querétaro (Research and Advanced Technology Centre of Querétaro)
CICESE	Centro de Investigación Científica y de Educación Superior de Ensenada (Scientific Research and Higher Education of Ensenada)
CIE	Centro de Investigación en Energía (Energy Research Centre)
CIEBT	Centro de Incubación de Empresas de Base Tecnológica, (Center for technology business incubation of the IPN)
CIMAT	Centro de Investigación en Matemáticas (Mathematics Research Centre)
CIT-UNAM	Centro de Innovación Tecnológica (Technology Innovation Centre)
CIMMYT	Centro de Investigación para Mejoramiento del Maíz y el Trigo (International Centre for Maize and Wheat)
CINVESTAV	Centro de Investigación y de Estudios Avanzados del IPN (Research Centre and Advanced Studies IPN)
CIO	Centro de Investigación en Óptica (Centre for Research in Optics)
CIQA	Centro de Investigación en Química Aplicada (Applied Research Chemistry Centre)
CITEDI	Centro de Investigación y Desarrollo de Tecnología Digital (Digital Research and Development Centre)
CIVAC	Ciudad Industrial del Valle de Cuernavaca, Morelos (Industrial City of Cuernavaca Valley, Morelos State Industrial Park)
COLPOS	Colegio de Postgraduados (The Postgraduate College, research center in agriculture research)
CONACYT	Consejo Nacional de Ciencia y Tecnología. (Science and Technology Council)
CONDUMEX	Centro de I&D de Condumex SA (Condumex Research and Development Center)
CUNITEC	Centro Universitario de Emprendedores Tecnológicos UdG,

	(Entrepreneurial Centre of the Guadalajara University)
Delta	Parque Industrial de Leon (Industrial Park in Leon Gto Mexico).
DF	Distrito Federal (Federal District)
ENESTYC	Encuesta Nacional del Empleo, Salarios, Tecnología y Capacitación en el Sector Manufacturero (National Census on Work, Salaries, Technology and Skill in the manufacturing sector)
FIDETEC	Fondo de Inversión y Desarrollo Tecnológico (Investment and Development Fund)
FIMEE	Facultad de Ingeniería Mecánica, Eléctrica y Electrónica, UGto (Mechanical, Electrical and Electronic Engineering Faculty, Guanajuato State University)
IB	Instituto de Biotecnologia, UNAM (Biotechnology Institute)
IEBT	Incubadora de Empresas de Base Tecnológica (Technology Base Incubator)
IETEC	Incubadora de Empresas de Innovación Tecnológica y Administrativa (Technology-based Firms and Business Incubator)
IIE	Instituto de Investigaciones Eléctricas (Electric Research Institute)
II-UNAM	Instituto de Ingeniería (Engineering Research Institute)
IMT	Instituto Mexicano del Transporte (National Institute of Transportation)
IMTA	Instituto Mexicano de Tecnología del Agua (Mexican Water Research Institute)
INDICO	Innovación Difusión y Competitividad (Innovation, diffusion and competitiveness, research project)
INIFAP	Instituto Nacional de Investigaciones Forestales, Agrícolas y Pecuarias (National Forest, Agriculture and Livestock Research Institute)
INCU-VEN	Incubadora del Programa de Vinculación de la UdGto (Incubator of the outreach program of the Guanajuato University)
INEGI	Instituto Nacional de Estadistica Geografía e Industria (National Institute of Geographical and Industrial Statistics)
IPN	Instituto Politécnico Nacional (National Politechnic Institute)
ITESM	Instituto Tecnológico de Estudios Superiores de Monterrey (Monterrey Institute of Technology)
LAPEM	Laboratorio de Pruebas de Equipos y Materiales (National Laboratory of electrical materials testing)
MABE	Centro de Tecnología y Desarrollo MABE (Research and

	Development Centre of MABE)
NAFIN	Nacional Financiera (National Financing Institution)
NRS	National Researchers System (SNI Sistema Nacional de Investigadores)
Pecyt	Programa Especial de Ciencia y Tecnología (Special program for science and technology)
PIEBT	Programa de Incubadoras de Empresas de Base Tecnológica de Conacyt (Technology-based Incubator Program of Conacyt)
PIEQ	Programa Incubador de Empresas de Querétaro (Firm incubator program of Querétaro)
PTA	Technology area of the research centers' projects
Provinc	Programa de Vinculación de Conacyt (Outreach Conacyt national program)
PyME	Pequeñas y Medianas Empresas (SME, Small and Medium Enterprise)
SEP	Secretaria de Educación Publica (Ministry of Public Education)
SIECYT	Sistema de incubación de empresas científicas y tecnológicas (Incubation system of scientific firms)
Torre de Ingeniería	Engineering Tower, multi-tenant building at UNAM
TREMEC	Transmisiones y Equipos Mecánicos S.A. de C.V. (Mechanical Transmission Equipment Company)
UAEM	Universidad Autónoma de Estado de México (State of Mexico University)
UAM	Universidad Autónoma Metropolitana (Metropolitan University)
UAQ	Universidad Autónoma de Querétaro (Querétaro State University)
UdG	Universidad de Guadalajara (University of Guadalajara)
UGCT	Unidades de Gestión de la Ciencia y la Tecnología (Technology management units in the RC)
UGto	Universidad de Guanajuato (Guanajuato State University)
UNAM	Universidad Nacional Autónoma de México (National University of Mexico)
UNITEC	Universidad Tecnológica de México (Technology University of Mexico)
VEN	Outreach program of the University of Guanajuato

Preface

As industrial economies are transforming into knowledge economies and newly industrialized nations are attempting to leapfrog into the knowledge era, interest in building knowledge regions has increased considerably. The special role of knowledge regions in the development of the innovative capacity needed to sustain global competition is now well recognized, regions rather than national economies being seen as the source of technical innovations. The success of knowledge regions as innovation poles not only depends upon local factors but on a host of national and international factors, which are equally instrumental in shaping this changing innovation-driven competitive landscape. The lessons derived from comparing regions recognized for their economic dynamism with regions less successful in their transformation may be considered as insightful by the policy makers, economic development officers, and other interested parties motivated to help shape policy for their regions.

The objective of this book is to analyze selected knowledge regions in the USA, Canada and Mexico, to derive lessons on the approaches, institutions and policies that are appropriate in one national environment and less so in another. The analysis focuses especially on the mechanisms used for nurturing innovative firms and fostering their agglomeration in each region. It builds on the theoretical work of past researchers and experience gained in various parts of the world in developing viable knowledge regions.

This book is the outcome of the 'Monarch' research project (Monarca in Spanish), a project focusing on technology incubation in knowledge regions in NAFTA countries. The Monarch butterfly was chosen as an icon for this project because, in spite of its apparent fragility, it finds the energy to fly over the three NAFTA countries during its short life. The Monarch butterfly stands for the power of knowledge as a major force in the development of NAFTA as the glue that binds the three countries together. The project research and data collection activities were carried out during 1998–2003. The Monarch project focuses on the science, technology and innovation capabilities of knowledge regions in NAFTA countries, on the opportunities for innovation-based regional development, and on the barriers that limit that development.

The USA, Canada and Mexico provide an interesting framework for this type of analysis because, even if they have solid trade relations and share the same continent, there are significant differences between the three countries in level of development, industrial infrastructure, education and systems of innovation, differences which may help to explain the characteristics, and finally successes and failures of our selected knowledge regions.

Linked by the NAFTA since its inception in 1994, the member nations together have 426 million inhabitants and produce more than $12 trillion worth of goods and services.[1] Canada and Mexico are the US's two top trade partners, followed by China and Japan. In 2003, US was the single largest trading partner for both Canada and Mexico, accounting for two thirds of foreign trade of each country. In the same year, US had a quarter of its total foreign trade with the two NAFTA partners – 15 per cent with Canada and 10 per cent with Mexico. The level of trade between Canada and Mexico is, however, limited; 4 per cent of the Mexican foreign trade is with Canada and only 2 per cent of Canadian foreign trade is with Mexico, though these figures have been increasing in recent years.

In 2002, the stock of US investment in Canada was estimated at $190 billion, compared with $170 billion for Canadian investment in the US. In the same year the stock of US investment in Mexico was roughly $90 billion (which is a significant 63 per cent of Mexican stock of foreign investment) and that of Canadian $5 billion (another 3 per cent of the Mexican stock of foreign investment). The stocks of Mexican investments in both US and Canada are not significant. NAFTA has also stimulated increased investment from outside countries, accounting for almost 24 per cent of global inward and 25 per cent of global outward Foreign Direct Investment.

By most accounts, NAFTA has created a very large free trade zone that has been beneficial to the three partners. It has, however, not solved the region's deficit in advanced technology products. Even if, among industrialized economies, the US innovation system is considered to be one of the most comprehensive and advanced (accounting for a major share of global research, development and innovation activities), the country's trade deficit in advanced technology products keeps on rising, reaching $27.4 billion in 2003. Canada and Mexico and the region as a whole also have growing trade deficits in advanced technology products. In terms of spending on R&D and innovation, the US comes first (spending 2.74 per cent of its GDP), Canada comes second (1.92 per cent), and Mexico is a distant third (0.39 per cent) according to 2001 figures. In a recent study, the World Bank has concluded that Mexico's deficiencies in education and research and development (R&D) limit the NAFTA power to enable the country to reach a better level of technological and economic progress.[2] The three countries must continue to develop their capacity for innovation, globally and at the regional level. In particular, learnings from their successful knowledge regions should be adapted to the less successful ones.

As outlined in the first chapter, the process that leads to the agglomeration of innovative firms in a region is complex. It involves a number of heterogeneous actors (technology-based firms, research centers, universities, governments, business services and so on). It requires extensive interactions, cooperation and exchanges between those actors; inter-organizational relationships and networking being the key to producing the organic environment that encourages creativity and entrepreneurial behaviors. The framework for analysis which is presented focuses on the analysis of those actors and their interrelations.

Fourteen knowledge regions or Technology Innovation Poles (TIP) are analyzed: four in the USA, four in Canada, six in Mexico (aggregated into four cases). The three country chapters that follow, one each for the USA, Canada and Mexico, start with an overview of each country's socio-economic characteristics and of its national system of innovation. They continue with the rationale for the selection of the regional cases selected for in-depth study, a short history of each case, an analysis of the elements that led to its development or hampered it, a description of the local innovation culture and incubating milieu and of some of its technology-based firms.

The final chapter provides a comparative analysis of the three countries and of the fourteen innovative regions that were analyzed, reflecting on regional success factors and constraints and ending with a summary of the lessons of the study focusing on potential technology development paths. It ends with an analysis of the potential for future cooperation in technological innovation between the three NAFTA countries, current cooperation being quite limited.

The three authors of this book, one from each of the three NAFTA countries, have directed research activities for the project: Sarfraz Mian for the US, Jérôme Doutriaux for Canada, and Leonel Corona for Mexico. The book is a result of a joint effort and the names are listed in alphabetical order.

During the research process, efforts were made to develop common questionnaires and interview guides for administration to the firms, research centers, science parks and incubators identified for the survey. In the implementation phase several challenges were faced including differences in definitions and interpretation, size and design of survey instruments, and ways to administer and collect data. However, wherever feasible a common methodology was employed for data collection and analysis. Occasionally, regional differences led to adjustments reflecting local realities. Incubating milieus meant science/technology/research parks and incubators/innovation centers in some cases and the whole regions acting as incubating spaces in others when there were no formal science parks and incubators present. The number of technology-based firms included in the US and Canadian case analyses were generally limited due to the low return rate attributed to survey fatigue experienced by many entrepreneurs in those countries. In Mexico, however, it was possible to administer a detailed survey to the firms as well as research centers through onsite visits. In most cases, the science park and incubator surveys were conducted through multiple site visits resulting in detailed interview data.

NOTES

1. In this book, $ is used for the US dollar, C$ for the Canadian dollar, and PPP$ for OECD's Purchasing Power Parity dollars, a currency estimate adjusted for local purchasing power.
2. Daniel Lederman, William F. Maloney and Luis Servén (2005a).

Acknowledgements

The Monarch project on which this book is based may not have been possible without the support of the Inter-Institutional Research Program for North America (PIERAN) of El Colegio de México, A.C. which provided initial funding. This project was one of the 26 research initiatives supported by this program between 1994 and 2004. The main goals of PIERAN are to support research on points of US–Canada–Mexico common interest, encourage collaborative research between US, Canadian and Mexican researchers, give visibility in Canada and the US to Mexican researchers, and train academic researchers (http://www.colmex.mx/centros/cei/pieran_call.htm, January 2005).

Additional support, for research expenditures and travel to conferences, was provided by the authors' academic institutions, the Faculty of Economics at Universidad Nacional Autónoma de México (UNAM), the School of Business at the State University of New York at Oswego, and the School of Management of the University of Ottawa. Support was also provided by UNAM through DGAPA research projects for part of the Mexican study (INDICO indices), for the Monarch project (IN 301602), and for the research team's conference travel (SIGHO-Conacyt) (Mexican Science and Technology Council, regional research fund). Also, the State University of New York at Oswego provided accommodation and workspace to the research team to help complete the book manuscript.

Partial or full financial support was received from the organizers of the various conferences where preliminary results were presented, including INFORMS meetings (in Atlanta, Montreal, San Antonio, and San Jose); Technology, Industry and Territory, International (TITI) seminars (in Querétaro, Pachuca, and Puebla); Second Congress of Americas (in Puebla); and Conference Board of Canada (in Montreal). Their support is really appreciated.

We believe that this project would not have been possible without the cooperation and support of the many science park and incubator managers and board members, regional development officers, public and university officials, university technology transfer administrators, technology entrepreneurs, academic researchers in the USA, Canada and Mexico, who agreed to be interviewed, sometimes more than once, and/or who completed a detailed survey. Confidentiality rules do not permit us to list their names and affiliations, but their proactive and generous support is warmly acknowledged. Additionally, a number of our graduate students and research assistants were involved in the literature review and data collection phases of this project, in the US, Canada and Mexico. Their hard work and dedication is hereby recognized. Finally, we are deeply indebted to our families, especially our spouses who made sacrifices on many occasions to allow us to be away to complete this project.

1. The key elements of innovation infrastructure and the evolution of knowledge regions: a framework for analysis

INTRODUCTION

The last quarter of the twentieth century has witnessed the unfolding of a fast paced innovation-driven global economy where knowledge and innovation are increasingly recognized as sources of global competitiveness and economic well-being. Scholars have shown that a country's capability to commercialize innovative products and services is related to its research activities and its proportion of scientists and engineers, and to its policies and programs supportive of research and its commercialization. It is also related to the development of geographically concentrated clusters of institutions and firms in a common field and to the quality of the linkages between those institutions and firms (Porter and Stern, 2002). The founding and growth of new firms for the creation of innovative products and services depends not only on the behavior of individual entrepreneurs, but also on the communities in which they live and work (Schoonhoven and Romanelli, 2002). Regions, rather than nation states, with their knowledge base, their innovative firms and their enterprising individuals, have been shown to be key contributors to innovation (Keeble and Wilkinson, 2000; Bresnaham and Gambardella, 2004).

The relationship between innovation, competitiveness and economic well-being has led to unprecedented efforts by policy makers at various levels: national, state, regional, and municipal to enhance innovation capability through policies that are placing an ever-increasing emphasis on collaborative research, on the effective development of new technology and on its speedy diffusion into the marketplace. As a result, a number of aspiring and knowledge endowed regions of the world have experienced a concomitant rise in their innovation and entrepreneurial activities. The infrastructure needed to generate an innovative business climate at the regional level includes policies encouraging research and investments, organizations and activities that enhance formal and informal networking, institutions of higher education and research centers, and often science or technology parks and technology business incubators for nurturing technology-based firms and growing firm clusters (Cooke, 2004).

The premise of this book is that to be successful in the twenty-first century innovation-driven global economy, we need to build sustainable knowledge regions. A knowledge region is defined as a territorial unit with abundant human and social capital, containing structures, organizations and people actively engaged in generating development through science, technology and innovation, and whose interactions achieve a high concentration of technology-based firms and highly skilled knowledge workers and entrepreneurs (Sanz, 2004). Past research on technical innovation and the commercialization of knowledge has shown the positive role of knowledge or learning regions, also known as regional innovation systems[1], with high-technology clusters acting as vibrant technology innovation poles (TIP) in the pursuit of technology diffusion (Cooke et al. 1998). Successful TIPs generally allow cohabitation of large anchor technology-oriented industrial organizations, small entrepreneurial firms, universities and research laboratories, business services and venture capital, good communication facilities, physical infrastructure and government services. They encourage cooperation between firms and research organizations, as well as competition, sharing expertise and knowledge in some domains, for example in the development of cooperative supplier networks, while specializing in specific niche areas (Saxenian, 1994).

Within the economic development and technology policy arena of the past three decades, there has been a parallel growth in the rise of new enterprise development tools such as technology business incubators[2] and science and technology parks.[3] As part of the regional innovation infrastructure, science parks and technology incubators can be viewed as mechanisms for promoting technological entrepreneurship through new venture creation and growth as they facilitate the flow of knowledge and expertise in a region between institutions of higher education, technology development centers and research laboratories. They serve to link technology, know-how, entrepreneurial talent and capital, and help to create and nurture the development of innovative technology-based firms (Smilor and Gill, 1986). Often established through collaboration among university, industry and government, they are intended to promote technology development and commercialization to the benefit of the local economy. These incubation mechanisms have been used in North America and elsewhere since the 1970s and are now considered integral parts of the innovation infrastructure of many regions, striving to replicate the conditions observed in the most successful US knowledge regions and high-tech clusters.[4]

To take into account the socio-economic context in which the innovation policies of the three NAFTA nations have evolved over time, we provide a comparative review of the three national innovation systems (NIS) (see relevant sections in Chapters 2, 3 and 4). From the comparison, it is obvious that the three innovation systems differ in multiple ways: the US having a comprehensive and most advanced NIS, Canada also having a highly advanced NIS, and Mexico in the process of developing one.[5] The technology business incubation practices in the three NAFTA countries are also somewhat different and diverse. While the US has a large number of well-established science parks and technology incubators spread across the nation, Canada has a smaller number of formal science parks and incubators, and Mexico has only recently embarked on experimenting with these tools.

This research effort focuses both on regions as systems of innovation and as incubating milieus and on property-based science parks and technology incubators as dedicated incubation mechanisms. Relatively more emphasis is placed on science parks and incubators in the USA; and on the regional innovation milieu in Canada and Mexico, since formal science parks and technology incubators have a less active role in those two countries, with weaker results in Mexico.

In order to pursue a comprehensive analytical approach to study knowledge regions in the three countries, we draw from the theory of the firm and its agglomeration, space-based regional view, and the evolutionary institutional perspective. Based on this integrative approach, we lay out our key theoretical arguments in favor of the knowledge region acting as technology innovation pole (TIP). First, the role of technology-based firms and their agglomeration as the core of innovation elements is established. Second, the evolutionary development of the TIP as innovation system is explained. Third, the key role of incubation mechanisms such as science parks and incubators is highlighted. Finally, a conceptual framework that includes all of the key elements of the regional innovation infrastructure is proposed (Figure 1.1).

1.1 AGGLOMERATION OF TECHNOLOGY-BASED FIRMS: THE CORE OF INNOVATION

Starting from the late 1970s, the role of technology and innovation in economic development came to the forefront after neo-Shumpeterian economists began recognizing technology as an important economic variable. For neoclassical economists, technology has been an independent variable in the sense that it is considered external to the economic system; that is, in the production function, technical change is taken as the residual that could not be explained directly by the key input factors, capital and labor. In fact, most economic theories simplify or confine technology to issues related to production: machinery, division of labor, production processes, product design and organization. Nevertheless, by the beginning of the 1980s most

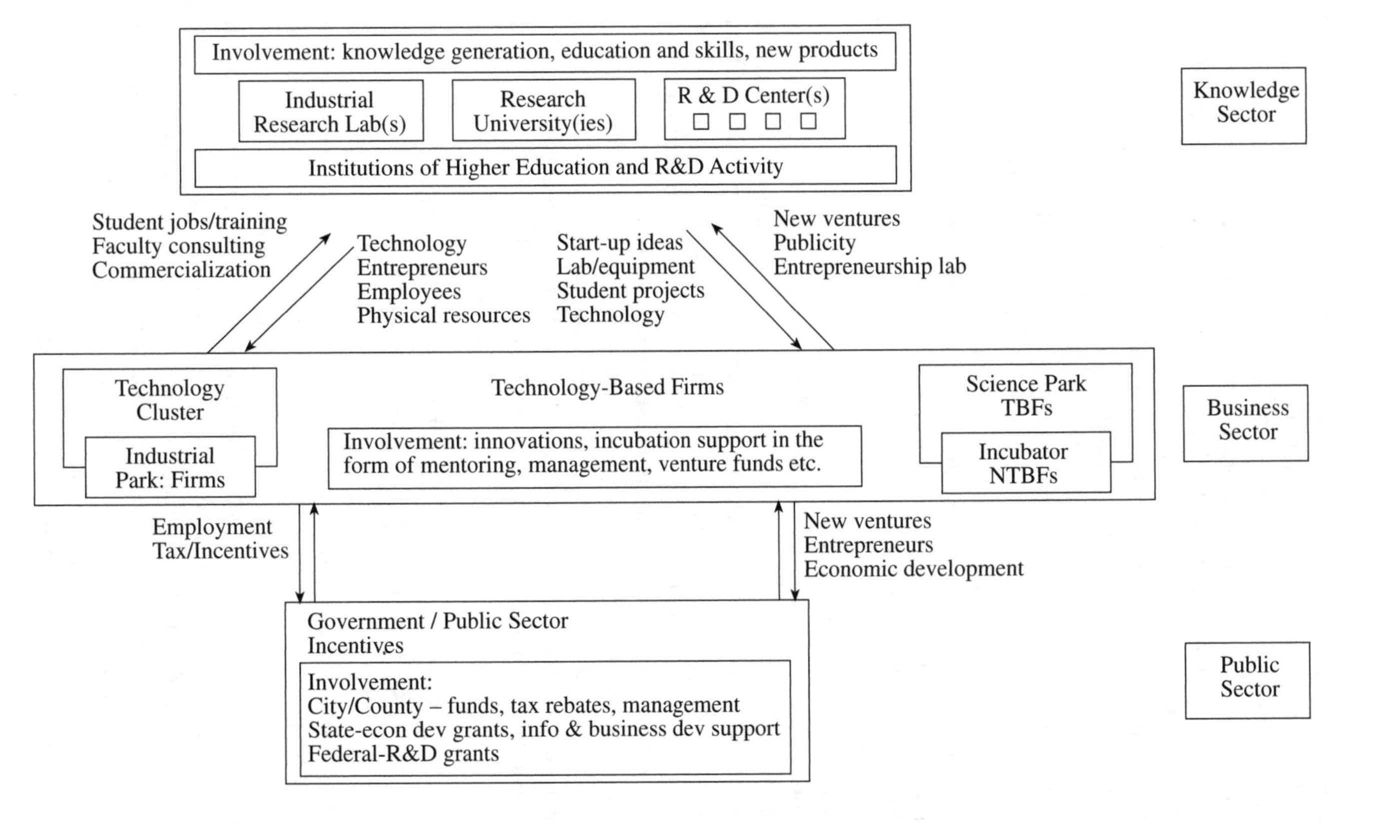

Figure 1.1 Key infrastructure elements of an innovation pole: a conceptual model

economists started to interpret technology as an endogenous variable to be explained by the economic conditions of production, which for neoclassical thought is already becoming the systematized theory of 'endogenous economic growth'[6].

There is now a broad conviction among economists that either as a dependent or an independent variable, technology should be considered as one of the main factors of development. Owing to the growing competition between persons, firms, regions, countries and blocks of countries, where technology has evolved to be a key driving force, it is both more important to identify the variables that explain technological change and to better understand the challenges posed by the increasing complexity of its analysis. However, as a general rule, any economic analysis of the components and impacts of technology depends upon its definition and measurement.

The core of the contemporary rate of technological change is related to the diffusion of technological product and process innovations that is 'introduced into the market (product innovation) or used within a production process (process innovation)' (OECD 1997a, p. 47). Therefore, the economic impact of these product and/or process innovations is based on the effectiveness of their diffusion, the depth and rate of penetration in the market. Technological product innovation includes both new and improved products. As defined by OECD (1997a, pp. 48, 49):

> A technologically improved product is an existing product whose performance has been significantly enhanced or upgraded. A simple product may be improved (in terms of better performance or lower cost) through use of higher performance components or materials, or a complex product, which consists of a number of integrated technical sub-systems, may be improved by partial changes to one of the sub-systems. A technologically new product is a product whose technological characteristics or intended uses differ significantly from those of previously produced products. Such innovations can involve radically new technologies, can be based on combining existing technologies in new uses, or can be derived from the use of new knowledge. Technological process innovation is the adoption of technologically new or significantly improved production methods, including methods of product delivery. These methods may involve changes in equipment or production organization, or a combination of these changes, and may be derived from the use of new knowledge.

Some of the technologies may be specific to a product or a service. They may also be 'platforms of technology' that can lead to new families of products, as in the case of the Windows operating system for micro-computing, the Radio Frequency Identification (RFID) for remote identification in telecommunications, or new genomics tools in biotechnology. And there are also 'disruptive technologies' that can lead to the creation of completely new industries, as in the case of the internal combustion engine or the internet, or improved production or managerial processes such as just-in-time manufacturing or electronic 'B2B' (Business-to-Business) communications.

Recent research on innovation has shown that at the national level, overall performance in technology and developing new products and processes, depends not only on the amount of research being done by a country and its regions (R&D expenditures, number of researchers), or on the measurable output of that research (number of patents and so on), but also to a great extent on the interactions between enterprises, institutions of higher education, and public and private research organizations. It further articulates that (OECD, 1997c, p. 9):

> Innovation and technical progress are the result of a complex set of relationships among actors producing, distributing and applying various kinds of knowledge. The innovative performance of a country depends to a large extent on how these actors relate to each other as elements of a collective system of knowledge creation and use as well as the technologies they use. These actors are primarily private enterprises, universities and public research institutes and the people within them. The linkages can take the form of joint research, personnel exchanges, cross-patenting, purchase of equipment and a variety of other channels.

Much has been written on the process that links science and technical innovation.[7] It has been shown (Roberts and Malone, 1996; Rothwell, 1994) that the process is inherently non-linear and involves individuals and organizations both in terms of knowledge flow to a specific technology or industry, and in terms of activities and their timing. Knowledge flows between certain areas of science and technology and industries, complementing the know-how of the productive sector, are described as 'innovation chains'; examples of such chains are microelectronics, information technologies, telecommunications, biotechnology, new materials, and new sources of energy. The activities and events occurring along the 'innovation chain' and their timing is described as a 'technological path', which includes the set of norms, standards, processes and procedures leading from all potential applications of a new technology to the solution of a specific market need, while meeting institutional socio-economic constraints. There are many potential technological paths associated with new innovation chains, involving technology-based firms (TBFs) and leading to various families of new products, processes or services. TBFs are, therefore, key players in the innovation process, which, with the help of public and private research institutes and universities, transforms science and technology into products or services with economic and market value.

Like many physical processes, innovations tend to follow a cycle. The 'innovation cycle' generally goes through three phases: the first phase is that of the introduction of a new product or service based on a new technology (product innovation); firm competitiveness comes from its 'first to market' status; entrepreneurial profits (extraordinary profit) tend to be high at this stage, but risks are also high. In the second phase, firm competitiveness comes from incremental innovations in the product/service, which becomes increasingly standardized. While in the third phase, firms must compete on price as the product and technology have reached maturity, so cost competitiveness depends on higher volumes, process innovations (organizational, operational and production issues) and other cost-reducing factors (such as offshore manufacturing to have access to cheaper labor or less restrictive environmental constraints).

In practice, product and process innovations are both present along the innovation–product life cycle, but as has been shown, in the third phase the relative importance of product innovations decreases in favor of process innovations (Davelaar, 1991). A positive institutional environment fostering innovations will generate higher profits (or higher expectations of profits) in the first two phases of the innovation cycle, and lower profits in the third.

Phases one and two of the innovation cycle benefit directly from 'agglomeration externalities' (Dorfman, 1983; Audretsch and Feldman, 1996) leading to self-sustaining industrial innovations in technology intensive regions where a number of conditions are present: a solid industrial infrastructure made of large and small firms, quality R&D activities, skilled labor, business services and financial resources, an entrepreneurial culture encouraging personnel mobility, social and professional networking and information exchange, supportive institutions working cooperatively, and a quality of life attractive to research professionals (Stuart and Sorenson, 2003). Phase three, on the other hand, often deals with 'standardized'[8] production, economies of scale, mass production, with competitiveness coming from the use of cheaper labor, the intensive use of energy and the exploitation of natural resources.

Typically, the economies of knowledge agglomeration, especially important in phase one and two of the innovation cycle, arise from one, two or all of the following conditions (Moulaeert and Djellal, 1995):

- A concentration of firms in certain industries (clusters)
- A regional infrastructure that provides a diversity of services, suppliers and promotional mechanisms that is mainly for innovation support
- Global and regional opportunities based on institutional networks and links among local suppliers that generate economic advantages for a firm.

Typical variables that facilitate or enable the agglomeration of knowledge are access to specific technical information and R&D personnel, links with research centers, an entrepreneurial milieu, a good supply of trained personnel, and programs and mechanisms (including incubators and science parks) that provide support to mediate the firm's higher risks of investments and spending on innovations.

1.2 THE KNOWLEDGE REGION AS TIP: CREATING A REGIONAL SYSTEM OF INNOVATION

The literature on regional technological capabilities holds that ways in which firms organize their economic activities reflect their institutional environment, and the ways in which people cooperate within the firm (Kogut, 1991). These knowledge capabilities are embedded in regional institutional settings. As a result, such regions and their nations can enjoy enduring leads in a particular industrial sector because a specific network of interacting firms and institutions is relatively favored at particular points in time. This has been a long-held belief (Marshall [1890], 1961) supported by numerous more recent studies (Biggart and Guillen, 1999; Almeida and Kogut, 1999) lending belief to the fact that it is important to look at the roles and specific practices of institutions in creating firm-level capabilities and organizing principles that are 'sticky', that is, reflect and adhere strongly to the contexts in which they were produced.

The heavily networked structures of present day technology-driven industries, and the heterogeneity of the actors and complex relationships within those networks, call for a framework that places technological accumulation and learning processes in the context of inter-organizational relationships. Researchers have proposed that technological innovation be studied in the context of a techno-economic network – a coordinated set of heterogeneous actors who interact to develop, produce and diffuse methods for generating goods and services (Callon, 1992). These networks are organized within an innovation-based region also called a knowledge pole or technology innovation pole (TIP). A knowledge pole or active TIP is an economic and geographic region where networks of institutions cooperate to create a positive milieu for innovation. TIPs are also sometimes called regional innovation systems (RIS): 'a regional innovation system consists of interacting knowledge generation and exploitation sub-systems linked to global, national and other regional systems for commercializing new knowledge' (Cooke et al., 2000).[9]

From an evolutionary point of view, the TIP can form as a result of the regional milieu in which market pull is the main force that creates an endogenous or spontaneous innovation milieu, and the region acts as a 'distributed research park' (see definitions in section 1.4). TIPs can also result from a deliberate policy decision with appropriate innovation mechanisms (science parks and incubators) established by public and/or private entities, and generally led by regional actors (exogenous models for innovation). In this case, technology push and the actors' intentions are the main forces for innovation. In practice, TIPs are often the outcome of a mix of endogenous and exogenous forces: which should be dominant and in which cases? These are important questions for policy makers to decide in light of each specific context.

Knowledge poles or TIPs offer a positive milieu for innovation through a set of interrelated institutions (Casas et al., 2000) where diffusion of innovation processes can take place in three geographic environments. The environment can be local or regional, with universities, TBFs and traditional industries, inter-related through networks of economic and knowledge relationships, and generating innovations in the form of chains in the region's areas of competitive advantage. It can be a national system of innovation characterized by national institutions. And it can also be in the form of much broader, international or global institutions providing additional inputs or even being the main sources of knowledge and resources. With different degrees of importance, the three environments are in play in each innovation pole (Krugman, 1991).

A typology of regional innovation systems, based on the type of governance observed in specific knowledge regions has also been proposed: (1) *grassroots*, locally organized; (2) *networked*, multi-level interactive, and (3) *dirigiste*, state dominated; each one has a different breadth of activity and potentially somewhat different criteria for success (Cooke, 2004). The role of governance in today's regional innovation systems is not of primary concern here and is beyond the scope of this book. The focus of this book is on TIPs in general and the factors that distinguish the successful from the less successful ones. This leads us to seek a better understanding of their developmental phases.

1.3 THE DEVELOPMENT OF THE TIP: AN EVOLUTIONARY APPROACH

The development phases of a typical TIP follow a standard evolutionary path: previous and emerging conditions, take-off, growth, maturity, stagnation and decline (Figure 1.2). Knowledge of these developmental phases in a TIP's agglomeration cycle provides insights for regional policy purposes. When trying to understand the factors that led to the emergence of a TIP in a region with a less-developed innovation infrastructure, it is useful to identify the

Cluster Stage ↓	Key Ingredients →	Endogenous (internal) Factors			Exogenous (External) Factors		
		Motivated entrepreneur	Opportunity recognition		Knowledge enterprise	Private industry/ business sector	Prospective regions & interested govt. entities
Previous & emerging		Champion entities			Champion entities		
Take off		Critical mass of entrepreneurs	One or more lead knowledge institution	Financial resources – Risk capital	Synergy of the exogenous factors: Creating an entrepreneurial environment		
			Research institute/ center, university		Incubators	Science parks	Other supportive programs
Growth		ACTIVE INSTITUTIONAL LINKS			ACTIVE INSTITUTIONAL LINKS		
Maturity		Inter-functional networks: in addition to performing its traditional functions, each institution assumes some role of the other			Inter-functional networks: in addition to performing its traditional functions, each institution assumes some role of the other		
Stagnation		Steady number of stagnating firms	Dysfunctional / Functional links				
Decline		Decreasing number of firms	Dysfunctional links				

Figure 1.2 The development of the TIP: an evolutionary approach

region's initial conditions (previous conditions in terms of location, urbanization, socio-economic base, and linkages). This is also useful when one is trying to understand the factors that support growth and create a milieu that is both attractive to TBFs and supportive of their development. The variables that facilitate the agglomeration of knowledge are expected to play an increasing role as the population of TBFs grow (the net result of local creations and in-migration of technology-based firms), until a 'critical mass' is reached. At that point there is a sustained supply of business opportunities with good potential coming from technology chains exploited in the TIP and there is enough local demand to sustain the technical, scientific and business services needed by industry (from molding, precision machining and prototyping to specialized financial services, intellectual property and corporate law firms).

This TIP development cycle can be compared with the phases of the Triple Helix development cycle proposed by Leydesdorff and Etzkowitz (1998):

a. *Take off*, when 'the three spheres are defined institutionally (university, industry and government). Interaction across otherwise defended boundaries is mediated by organizations such as industrial liaison, technological transfer, and contract offices'.
b. *Growth*, when 'the helices are defined as different communication systems consisting of the operation of markets, technological innovations … and control at the interfaces. … The interfaces among these different functions operate in distributed modes that produce potentially new forms of communication as in a sustained technology transfer interface or in the case of patent legislation'.
c. *Maturity*, when 'the institutional spheres of university, industry, and government, in addition to performing their traditional functions, each assume the role of the others, with universities creating an industrial penumbra, or performing a quasi-governmental role as a regional or local innovation organizer'.

These phases can be operationalized by exploring the following main aspects in the actual TIP studies presented in this book (Figure 1.2; see also section 5.2):

- Pre-conditions: short description of the socio-economic characteristics of the area at the beginning of its potential transformation into a knowledge pole, with a special focus on factors such as: local research orientation, universities, level of education, availability of trained manpower, main economic base, quality of life.
- Take-off: outline of the conditions/events that led to the decision to develop a knowledge base, lead actors, main infrastructure developed, major milestones.

- Current knowledge-base clusters: area's strength; targeted sectors.
- Current infrastructure: higher education, major research laboratories, main TBFs, formal incubation infrastructure, venture capital and other sources of financing.
- Overall assessment of actual conditions in terms of the area's performance as a competitive and sustainable knowledge pole, threats and challenges.

The growth of an existing innovation pole or the emergence of a new pole can be attributed either to the development of a new technological trajectory, a chain of different economic sectors that are part of product or process innovations, or the renewal of traditional or stagnant industries. These sources can have different levels of technology: high, medium or low, leading to different kinds of innovations depending on market and technological flexibility, in inverse relationship with the degree of standardization of products and processes at play. TBFs can be based on new technologies, new industrial processes, or they can lead to the renewal of existing industrial sectors. TIP sustainability depends upon the economic conditions for agglomeration of TBFs and its ability to reach a 'critical mass', first in specific industries and, second, in institutional milieus for innovation.[10]

In some cases, formal incubation mechanisms are employed to assist in the enterprise-building process and to nurture emerging TBFs, which brings us to the next section.

1.4 SCIENCE PARKS, INCUBATORS AND THEIR EVOLUTION: PROVIDING MECHANISMS FOR NURTURING INNOVATIVE FIRMS

Technology incubation mechanisms are known under various names and definitions – technology incubators, innovation centers, science parks, and technology centers. Their common features are the provision of physical space and intangible support to new firms engaged in the development and commercialization of knowledge-intensive products or services, and, in most cases, their links with public or private research laboratories or research universities (OECD 1997b, p. 5). The advent of 'incubators without walls' or virtual incubators in the late 1980s and of regions acting as distributed parks has now made the provision of physical space optional, intangible support being the key characteristic of today's incubation programs. This leads to the concept of 'incubation space' to describe all the physical and virtual spaces in which business incubation takes place.

One of the foci of this study is on both science/research parks and on technology incubators as elements of an 'incubation space'. A science park will generally be expected to be a land-based campus-like development designed to accommodate technology-based firms, research institutes, and appropriate support services. A technology incubator is an organization providing intangible services to new technology-based firms (TBFs), technology entrepreneurs, and technology spin-offs to facilitate their start-up and early development.

Science parks are expected to create a 'critical mass' of technology-based activities providing member organizations with an environment conducive to the cross-fertilization of technical and business ideas and supportive of the commercialization process, thus contributing to regional economic development and job creation. The environment is designed for research and development organizations, high-tech firms and support services providers, and is often linked with a university. Although the first formal science park was established in 1951 at Stanford University, science park creation really took off in the 1980s in an effort to support economic development and job creation, and to encourage the growth of knowledge-based industries. In the US, 91 science parks were established in the 1980s compared with 32 parks between 1951 and 1980. In Canada, 12 parks were created in the 1980s while only three parks were founded in the 1960s and 1970s. Mexico's experience with science parks is relatively recent (1990s) and limited. According to the Association of University Research Parks (AURP)[11] there are now over 410 such parks at various stages of development in the world, out of which 136 are in the US.

The science/research/technology parks form a very heterogeneous population. For the purpose of this analysis, the following three groups are identified:

1. Industrial Research Parks; campus-like developments generally created by land developers and/or regional governments to attract large corporations looking for an attractive area to locate their research facilities. These parks tend primarily to be real estate developments with plenty of green space, designed for scientific or industrial research laboratories. Services offered to park tenants tend to be limited; inter-firm linkages and relationships with universities and/or government laboratories being left to the discretion of each firm.
2. University-Related Research Parks; campus-like developments created by universities or by local governments with strong university input. Their initial objectives are generally to be university income-generating ventures and to become regional development growth poles, and also to be a mechanism for the commercialization of university technology through spin-offs.

3. Distributed Research Parks, with no physical location, when a city ('Technopole') or a complete region behaves as a science park. Research activities, small and large high-technology companies, universities and technical colleges, technology transfer services, support services for start-ups, business services, are all distributed geographically over the region, the complete region acting as a giant science/research park.

Since its first inception in the USA in the 1950s, the concept of business incubation has evolved into a diverse and heterogeneous industry. For many years, incubators were strictly property-based developments offering temporary space and business services to start-ups to help them during the initial phases of their development. The concept of the incubator has now evolved to include a wide variety of programs and activities designed to encourage and support the development of start-ups and growing firms. According to the National Business Incubation Association (NBIA)[12]

> Business incubation is a dynamic process of business enterprise development. Incubators nurture young firms, helping them to survive and grow during the start-up period when they are most vulnerable. Incubators provide hands-on management assistance, access to financing and orchestrated exposure to critical business or technical support services. Most also offer entrepreneurial firms shared office services, access to equipment, flexible leases and expandable space all under one roof.

NBIA estimates that in 2003 there were 950 incubators in North America, compared with 587 in 1998, and only 12 in 1980.

A high percentage of incubators and incubator programs are not-for-profit ventures, public or private, with, most often, job creation or economic diversification as their primary objective, technology transfer and technology commercialization being another common objective, and the empowerment of under-privileged groups also sometimes being an aim. They may be 'mixed use' in an urban, suburban or rural setting with no specific sectoral orientation (47 per cent of North American incubators according to NBIA), focused on technology start-ups and growing firms (37 per cent) and, in that case, often located in a research park, on a university campus or next to a large research laboratory, or they may be more specialized. Technology incubators, generally sponsored by universities, governments, economic development agencies are more likely to be subsidized. They provide management assistance and business counseling to start-ups, as well as access to business services, access to laboratories and R&D, networking, and assistance in finding external funding. Some incubators are business investments with a clear for-profit orientation designed to nurture and grow technology-based ventures. They represent a small percentage of the total number of incubators and are generally created by large corporations, professional groups, large consulting firms, and venture capital groups. The number of 'for-profit' incubators, estimated at 350 in North America in 2000 at the height of the 'dot-com bubble' has decreased significantly in the early 2000s.[13]

Until the mid-1980s, business and technology incubation took place in formal 'brick and mortar' incubators, which were multi-tenant buildings with flexible rental space, common facilities, management assistance and business counseling, and affordable business services designed specifically for start-up firms. Some incubators then started to provide management assistance and business counseling to non-tenant 'affiliate firms', and other incubators were created directly 'without walls', incubation programs offering active management support, business counseling and networking opportunities to start-ups without the provision of rental space. They are sometimes called 'virtual' incubators, a name also applied to incubation programs making extensive use of the internet for the delivery of their services.

The late 1990s saw the appearance of new forms of incubator organizations to nurture and grow private technology companies, often referred to as 'technology incubators': the 'hubs', the most common for-profit mode, consisting of a central office with specialized divisions offering specialized services to member start-ups (accounting, marketing, IP, hiring, purchasing and so on); the 'networked incubators', with a 'strong core ensuring good information flow, and multi-layered orbits/affiliations' to encourage collaborations and partnerships among members and graduates and to have key services offered by members within the network; the 'econets', which were 'operating companies [providing incubation services] that buy majority stakes in start-ups and pool them into groups that allow them to leverage off each other', and 'metacompanies', or businesses that produce companies. The appeal of those 'for-profit' models has decreased considerably with the high-tech meltdown of the early 2000s (Hansen et al., 2000, pp. 74–84).

A new form of technology incubator which started in the early 2000s is the 'accelerator'. Those post-start-up incubators are designed to help young technology-based firms achieve a high rate of growth. There is also currently an increasing trend towards niche incubators, and an equally increasing trend towards partnerships and alliances between incubators, and within incubators to benefit from stronger networks and organic relationships.

In summary, within the past quarter of a century both the research park and incubator industries have grown rapidly and the various models have matured considerably. The heyday of the park incubator movement was during the last two decades of the twentieth century in which it prospered during growth spurts and survived in economic downturns. In future, given the generally positive experience in the US and Canada, and mixed results in Mexico, it may well be that every community that aspires to establish knowledge regions will consider making use of these tools in accordance with their local needs.

1.5 A FRAMEWORK FOR ASSESSMENT: BENCHMARKING THE TIP

As in any other systems of governance, benchmarking may serve as an important concept for assessing the performance of emerging knowledge regions. Comparisons between TIPs are complicated by wide variations in regional models, country-specific variation in modality, and variations in local environments and resource endowments. These contextual elements often limit the scope of comparisons but still allow benchmarking to be used to identify examples of best practices.

Based on the above analysis and synthesis of the regional innovation context, the various elements of the TIP innovation infrastructure are organized under four headings: the key regional actors, the context, the innovation process enablers, and the outcomes. They are described below:

a. Key regional actors
- *Technology-based firms* – technology-based firms (TBFs) and their complementary industrial activities: mix of large firms and of small entrepreneurial firms
- *Government programs and incentives* – committed public entities at various levels of government (national/federal, state/province, regional/local) with supportive policies and programs designed to promote and sustain innovation
- *Universities and R&D centers* – a solid technological knowledge-base embedded in a research infrastructure made of public and private research laboratories, research universities, and their R&D units.

b. The regional context
- *Entrepreneurial culture* – a local culture supportive of entrepreneurial activities, with successful role models and cooperation among local stakeholders to promote such activities
- *Qualified manpower* – a good supply of human resources at various levels of skills and qualifications, and the institutions (universities, post-secondary technical institutions) needed to provide that training
- *Quality of life* – provision of attractive quality of life amenities to researchers, professionals and their families
- *Lower cost of doing business* – availability of low cost labor (graduate students, faculty consultants and so on), low cost utilities, affordable cost of living, low raw material and transaction costs.
- *Traditional industrial base* – presence of a developed industrial base and related services as source of raw materials and markets for exchange

of goods and services and as source of basic services

- *Regional infrastructure* – presence of modern communication and transportation facilities such as broadband telecommunications, airport with direct links to other important TIPs, markets, and so on)

c. The innovation process enablers

- *Formal incubators* – technology business incubators, innovation centers, business technology centers
- *Science parks* – science / technology / research parks
- *Financial resources* – risk capital in the form of seed and venture capital, government grants and funding programs
- *Champion entity* – champion entities in the form of committed individuals and/or organizations providing leadership roles
- *Anchor organizations* – large technology firms, research laboratories or their units, and other support organizations. These organizations generally acquire early tenancy with longer-term commitment, providing operational stability to the incubation mechanism
- *Technology transfer programs* – presence of significant university–industry, research laboratory–industry or related technology transfer activity
- *Support services* – The technical, engineering, business, legal (including intellectual property) services needed by TBFs, and other services
- *Networking opportunities* – encouraging formal and informal networking in the region, providing forums for exchanges and cooperation among local stakeholders, facilitating technology transfer.

d. Outcome

Outcome in the form of regional development and sustainability measured by:

- *TBFs and firm clusters* – a sustained flow of newly created and in-migrating TBFs, a high level of retention and sustained growth leading to solid TBF clusters with a good critical mass in selected domains.
- *Innovations and new products* – a sustained output of innovative products, services and processes

This assessment framework highlights the key drivers that are expected to influence the extent to which TIPs develop, grow, and become self-sustainable; the extent to which TIPs provide a 'seamless continuum' for regional innovation (see Figure 1.2). This framework includes characteristics of TIP performance that can be benchmarked and is illustrated through the cases analyzed in this book.

1.6 THE RESEARCH PROCESS AND ORGANIZATION OF THIS BOOK

As outlined in the previous sections, the focus of this book is on technology-based firms and their agglomeration in technology innovation poles, and on the mechanisms used for nurturing innovative firms. The project started with the development of a framework for the analysis of technology incubation infrastructure and activities at the regional level and of the incubated firms (this chapter), and the preparation of data collection instruments in English, French and Spanish, the three official languages of the NAFTA countries. Data collection was then performed in each country/region, leading to specific country studies and to a three-country comparative analysis.

This book continues with the country chapters, one each for the USA (Chapter 2), Canada (Chapter 3) and Mexico (Chapter 4), and employs the framework proposed in Chapter 1. Each country chapter starts with a socio-economic overview followed by the national and regional system of innovation. The chapters then provide in depth analyses of the selected innovation poles in each of the three countries, and explores characteristics of the local incubating milieus/mechanisms and their technology-based firms. The population of interest includes 11 Canadian knowledge poles, 15 US knowledge poles with a major science park and six main Mexican knowledge regions/poles. The process that led to the identification of that population and to the selection of the four US, four Canadian and six Mexican cases is outlined in Chapters 2, 3 and 4 (Figure 1.3).

The comparative analysis of the three countries and of their innovative regions come in the final chapter of this book. The chapter focuses on the fourteen regions grouped in twelve cases covered in the study, four in each country, reflecting on the success factors and the constraints that tend to be associated with the more successful and the less successful regions. It summarizes the lessons of the study, reaffirms the benchmarking elements of the TIP infrastructure proposed earlier along with the potential technology development paths of knowledge regions, and analyzes the potential for cooperation in technological innovation between the three NAFTA countries. Because of the variability of national innovation systems, regional infrastructures and incubation mechanisms, the country chapters are somewhat different for each country, with relatively more emphasis on science parks and incubators as incubating spaces in the USA, and on the regional milieu as larger incubating space in Canada and Mexico. Whereas US TBFs included in the study were located in either a science park or its technology incubator, TBFs surveyed in Canada were generally located in a formal incubator (only in Saskatchewan did the study include firms in a science park outside an incubator), and TBFs surveyed in Mexico were located either in an

Figure 1.3 Top North American technology innovation poles

incubator, in a science park or in the region outside the park. Limited research funding and local constraints also led to significant differences in TBF response rates to the surveys done for this project, with the participation rate in Mexico being the highest followed by the US and Canada, where results were generally affected by survey fatigue among TBFs in incubators and science parks. Those differences limited the depth of analysis of the inter-regional comparisons and the learnings from a comparison of TBFs located in different national and regional systems of innovations and different socio-economic settings.

NOTES

1. The phrase 'regional innovation system' (RIS) is often used to describe a large geographic region, a state, a province. In this book the words 'region', 'pole' and 'area' will be used as synonyms to describe a metropolitan area, an area small enough to allow easy personal interaction with no more than an hour or two of driving time.
2. A technology incubator is 'a property-based venture which provides tangible and intangible services to new technology-based firms, entrepreneurs, and spin-offs of universities and large firms, all with the aim of helping them increase their chances of survival and generate wealth and jobs and diffuse technology.' (OECD, 1997b, p. 2)
3. A science park (popularly known as research parks or technology parks in the US and Canada) is a 'property-based initiative which has operational links with universities, research centers and other Institutions of higher education, is designed to encourage the formation and growth of knowledge-based industries or high value-added tertiary firms, normally resident on site, and has a steady management team actively engaged in fostering the transfer of technology and business expertise to tenant organizations' (OECD, 1997b, p. 2). The main objective of science parks is to encourage the collaboration of research and design functions and technology know-how, thus leveraging real and intellectual capitals.
4. For a more detailed analysis see Mian (1997).
5. Lederman, et al. (2005b). http://www.sice.oas.org/geograph/north/le_ma.pdf, January 2003.
6. Economic theories have increasingly been including explanations of technological change. Neoclassical theory explicitly includes technology and other related factors (such as education) as part of the production function, the so-called endogenous growth theories. Thanks to the pioneering works of Romer (1986) and Lukas (1988) such efforts in endogenizing the treatment of technology have started to impact mainstream economic literature in a short period of time (see Aghion and Howitt, 1997).
7. For details see OECD (2001c), OECD (2002a), OECD (2002b), OECD (2003a).
8. A *standardized* product is made with known, widely diffused production technology in which quality is so widely attainable that competition comes to be inevitably centered on price, and mainly organized by economies of scale (for details see Storper, 1997, p. 109).
9. Successful TIPs have a dynamic regional innovation system with solid research institutions, TBFs to commercialize that research, a culture that encourages networking, collaboration and entrepreneurship, and services and an infrastructure supportive of business activities. They are also recognized by a quality of life which makes them attractive to highly skilled professionals (for a more detailed analysis see Doutriaux, 1998).
10. In explaining innovations, industry comes first, and the region second. An alternative proposition is the one that focuses 'on innovation and the knowledge infrastructure and not on specific industries, a technology, or institutional reform' (Leydesdorff and Etzkowitz, 1998, p. 201).
11. Association of University Research Parks (AURP), Reston, VA. USA; http://www.aurp.net.
12. National Business Incubation Association, Ohio, USA; http://www.nbia.org/.
13. Forecast in 2000 of a decrease from 350 to 100 by Hansen et al. (2000, pp. 74–84) and even to 50 by NBIA.

2. US technology infrastructure and the development of regional innovation poles through incubation mechanisms

INTRODUCTION

The United States is a leader in the development of knowledge regions and in enhancing their capacity to create, attract and expand innovation-producing assets to leverage them for carving a 'new economy' space. To participate more fully in this innovation-driven 'new economy' many US regions, states and localities have striven to emulate the success of the legendary technology innovation poles (TIPs) – Silicon Valley and Boston's Route 128 – with the development of new and emerging TIPs, the focus of this book. In doing so, they have reinforced the local knowledge infrastructure by building institutions of higher education and R&D organizations; they have forged partnerships and designed policies supportive of networking; and they have often relied on modern technology and enterprise development tools such as science parks and incubators to nurture technology-based firms.

To better understand the context of innovation, this chapter starts with a review of the US national innovation system at federal, state and regional levels. This is followed by a discussion on the representative cases of knowledge regions or TIPs selected for study. The in-depth analysis of each of the selected cases explores the history of their development, their current status as technology innovation poles, and studies the main science and technology park facilities that were established as focal points to develop the knowledge region. A common format is employed for each case study, and the chapter ends with a summary and conclusion.

2.1 THE US NATIONAL INNOVATION SYSTEM

Recently, the US Council on Competitiveness, a group of prominent corporate chief executives, university presidents and labor leaders, resolved:[1] 'Innovation will be the single most important factor in determining America's success through the 21st century. For the past 25 years, we have optimized our organizations for efficiency and quality. Over the next quarter century, we must optimize our entire society for innovation.'

The truth of this statement is evident from the fact that today, among the large high-income industrialized countries of the world, the US innovation system is considered highly productive by most standards. The US innovation system has several salient features, which set it apart from those of other nations. First, among the industrialized economies, the US innovation system is one of the most complex, well established, and extensive enterprise, accounting for a major share of global research, development and innovation activities (National Science Board, 2004). The US national innovation policy is multifaceted, ingrained and wide ranging. Second, the relative importance of three key sectors within the US innovation system – university, industry and government – as performers and as funders of R&D also contrasts somewhat with the role of these sectors in many other national innovation systems, although their roles within the US system itself have evolved considerably during the past several decades. Third, a key characteristic of the US innovation systems (which is particularly noteworthy for the post World War II period and became more prominent after the end of the Cold War) is the importance placed on small and medium sized enterprises (SMEs) in the commercialization of new technologies developed during the intensive defense R&D periods (Mowery and Rosenberg, 1993). As a result, in the US, small and emerging technology-based start-ups have played a significant role in the development and diffusion of microelectronics, computer hardware and software, biotechnology, robotics and so on during the past five decades.[2]

This study of the US National Innovation System is divided into three sections. First, an overview of the system's socio-economic and science and technology context is provided in light of its historic development and current performance. Second, a discussion of the federal system of innovation highlights the critical role played by the national government. Third, the states and regional systems of innovation are discussed to elucidate the decentralized nature of the system as detailed below.

2.1.1 The Socio-Economic and Science and Technology Context

The socio-economic overview

Located in the center of the North American continent between Canada in the North and Mexico in the South, mainland America along with Alaska to the North West of Canada, and Hawaii in the Central Pacific Ocean, make up the 50 states of the United States of America. It covers an area of 3.5 million square miles (9 million square kilometers) of which around 5 per cent is made up of water (comprising the Great Lakes, inland and coastal water).[3]

The US population according to the 2003 estimates is almost 291 million.

During the past decade (1990–2001) the population increased at an annual average rate of 1.2 per cent.[4]

The country is divided into four geographic regions comprised of nine divisions. These are: the Northeast Region comprising the New England and Middle Atlantic divisions (population 54 million i.e. 19 per cent); the Midwest Region, comprising East North Central and West North Central divisions (population 64 million i.e. 23 per cent); the South Region, comprising the South Atlantic and East South Central division (population over 100 million i.e. 36 per cent); and the West Region, comprising the Mountain and Pacific divisions (population over 63 million i.e. 22 per cent).[5]

According to the latest estimates (OECD, 2004c), the nation's gross domestic product (GDP) was PPP $10 383 billion in 2002, which is $36 120 per capita. While the GDP per head increased, in real terms, by an average of 1.9 per cent per year, the overall real GDP growth showed an annual rate of 3.1 per cent during 1990-2001 – however, in 2001 it was 5.0 per cent, and in 2002 it slowed down to only 0.3 per cent.

Agriculture (including forestry and fishing), with 2.4 per cent of the working population employed in this sector contributed 1.4 per cent of GDP in 2001. Industry (including mining, manufacturing, construction and utilities), with 22.6 per cent of the civilian working population employed in this sector contributed 22.2 per cent of GDP in 2001. Services provided 76.4 per cent of GDP and 75.0 per cent of total civilian employment in 2001. The combined GDP of all service sectors rose in real terms at an average rate of 4.3 per cent per year during 1994–2001. Services GDP increased by 5.2 per cent in 2000 and by 2.1 per cent in 2001.[6]

Education is primarily the responsibility of state and local governments, and is free in every state from elementary school through high school. At the beginning of the 2001–2002 academic year, there were 33.6 million pupils enroled in public primary schools and 13.6 million in public secondary schools. Private school enrollment was approximately 4.6 million at primary level and 1.3 million at secondary level. There were almost 4000 two-year and four-year colleges and universities, with a total enrollment of about 15.3 million students at the beginning of the 2001–2002 academic year. Federal government expenditure on education (including training and employment programs) totaled approximately $80 billion in 2002. Spending on education at all levels of government in 1999–2000 was $521.6 billion (roughly 30 per cent of total public expenditure).[7]

The percentage of a nation's business base classified as high technology provides a measure of the extent to which the nation's industrial base is poised to capitalize on new technology and innovation. These industries include both manufacturing and service industries where technology is rapidly evolving. As the US national economy shifts towards high value-added products and services, the regions with the highest percentage of high-technology business establishments will be best positioned to take advantage of this shift. In 2000, there were 425 992 establishments in the US that were classified in the 39 high-technology NAICS codes[8] (Table 2.1). This represents 6 per cent of the more than 7 million total establishments in all 50 states in 2000. A study by the Milken Institute (1999)[9] shows that the gross industry output for the US high-technology industries has been on a steady increase over the last three decades, accounting for almost 11 per cent of the total US industrial output in 1997. The demise of the far-fetched internet businesses and the resulting high technology market slowdown in year 2000 proved that the high-tech sector has undergone a cyclical downturn, if not structural adjustment. The US high-technology sector has been in a consolidation phase since then.

Brief history of science and technology

The expansion of the American economy during the late nineteenth and early twentieth centuries combined with innovations in transportation, communications, and production technologies yielded manufacturing operations on an unprecedented scale (Chandler, 1997). Production activities, built on a long established pattern of technological innovation and adaptation, relied largely on mechanical skills. Growth in manufacturing productivity and output in the nineteenth century US economy was achieved in part through the development of the 'American system of manufactures' for the production of light machines and other mechanical devices. Therefore, innovations in this sector did not rely heavily on scientific research. (David, 1975; Rosenberg, 1972).

The resource endowment of the US, which favored the development of machinery for agriculture and transportation applications, its enormous protected domestic market, and the ability of the US to exploit foreign sources

Table 2.1 Technology intensity of the US industrial base (2000)

Firm characteristic	Total industry	High technology industry	High tech as % of the total
Number of firms	7 050 393	425 992	6.0
Number of employees	113 649 993	10 050 578	8.8
Employee payroll	$3 859 445 547	$583 276 953	15.1
Births of new firms	707 251	55 535	7.9
Net firms formations (birth-deaths)	N.A.	9 663	13.7 per 10 000 firms

Source: Office of Technology Assessment, US Dept of Commerce, 2004

of knowledge (importing machinery, blueprints, and skilled tinkerers from around the world) all supported these developments.

Much of the structure of the private sector components of the US national innovation system was established during the 1900–1940 period. The chemical, rubber, petroleum, and electrical machinery industries were among the most research-intensive, accounting for more than half of the total employment of scientists and engineers in industrial research within manufacturing, throughout this period. Five states (New York, New Jersey, Pennsylvania, Ohio and Illinois) contained more than 70 per cent of the professionals employed in this research effort that contributed significantly to the stability and survival of the giant diversified corporation that began at the turn of the century. One exception to the pattern of stability in research intensity during this period is transportation equipment, owing to the rapid growth in automobile and defense-related aircraft industries. Prior to 1940, the level of federal and state funding for industrial research was limited; the government financial support was primarily confined to the industrial extension program and to aid the wartime research effort in electrical machinery and instruments. Similarly, university research budgets before 1940 were also limited; the university system was one in which the requirements of industry, agriculture, and mining were recognized and accommodated. As research within industrial establishments grew in importance and volume, university research during this period often involved various forms of collaboration with private industry.

Figure 2.1 outlines the post World War II evolution of the US Technology Policy that has shaped its current innovation system.[10] In the first quarter century following the war, the United States enjoyed global competitive and technological dominance. Many of the path-breaking technological innovations occurred in the US, and as a result the country did not experience significant competitive challenges from abroad.

During the first post World War II period (1950–1980), the federal government's science and technology policy emphasis was on basic research and government mission oriented R&D. The government made especially large investments in defense and space-related research and development in response to the Cold War and race in space against the Soviet Union. During the more recent emerging information and globalization era (1980–2000), the emphasis has shifted to developing partnerships among various sectors of the national economy. Through these partnerships, the federal government has developed more active cooperative policies. The new paradigm for the development and deployment of technology that emerged has revolutionized the roles of public and private sectors alike (Alic, 2001; Bergland and Colburn, 1995). The details of these programmatic activities are provided in the following sections.

	WWII	1970	1990 2010
Dominant world policy	Cold War		Global Economic Growth
World competitive situation	US Dominant	Triadic: US, Europe, Japan (Relative US decline)	Global US Resurgence?
US Government S&T policy	Mission • Defense • Space • Energy • Health • Basic Research		Spin -Off (Slow commercialization)
Civilian competitiveness		1980 – University/Government/ Industry partnerships (faster commercialization)	2000 – Emphasis on TBED by building knowledge regions, high-tech cluster

Source: US Department of Commerce, 1997 (with authors' elaborations)

Figure 2.1 Evolution of the US science and technology policy

Science and technology indicators

According to official data the US Gross Domestic Expenditure on R&D (GERD) was $284.6 billion in 2003, which is 2.62 per cent of the US national GDP.[11]

As shown in Table 2.2 (a), over the past five decades, the US GERD has increased gradually in current dollar value, with a gradual increasing trend from 2.27 per cent to 2.72 per cent over the past four decades. In 2000, around $26 billion of US R&D funds (almost 10 per cent of US GERD) have come from foreign sources – a 124 per cent increase in ten years.[12]

Table 2.2 (b) shows the trends in the share of national R&D expenditures by funding sector over the past five decades. While the numbers show a steady increase in all categories of funding sources, the higher growth rates in industry funding and the flattening of federal funding growth rate are significant beyond the 1980s. The data show an increasing share (beyond 60 per cent) of industry dollars since the mid-1990s which has vastly improved the US innovation system.

Table 2.2 (c) details national R&D expenditures by sources of funds and character of work. As depicted, the federal government provides most of the funds (59 per cent) used for basic research, while industry supports most of the development (81 per cent) as well as applied research funds (61.5 per cent).

Table 2.2 US science and technology funding statistics

(a) Overall

	1960	1970	1980	1990	2000	2003
GERD, $ billion (current)	13.7	26.3	63.3	152.0	265.2	284.6
GERD/GDP, per cent	2.60	2.53	2.27	2.64	2.72	2.62
Foreign funds $ billion (current)	n.a.	n.a.	1.9	11.5	25.8	n.a.

Source: OECD Main Science and Technology Indicators – 2004/1

(b) By funding sector (per cent)

	1960	1970	1980	1990	2000	2002
Federal government	65.0	57.0	47.5	40.6	26.3	28.0
Industry	32.9	39.8	48.8	54.7	68.4	66.0
Higher education	0.5	1.0	1.5	2.1	2.3	3.0
Non-profit & other	1.6	2.1	2.2	2.6	3.0	3.0

Source: National Science Board – Science and Engineering Indicators, 2004

(c) By source of funds and character of work in 2002 (per cent)

	Source of funds			
	Federal government	Private industry	Universities and colleges	Other non-profit
Development	17.6	81.4	0.3	0.7
Applied research	31.7	61.5	3.0	2.9
Basic research	58.9	18.5	10.2	8.9

Source: National Science Board – Science and Engineering Indicators, 2004

Table 2.2 Continued overleaf

Table 2.2 Continued

(d) By performing sector and character of work in 2002 (per cent)

	Performing sector			
	Federal* government	Private industry	Universities and colleges	Other non-profit
Development	9.0	89.0	0.8	1.0
Applied research	15.9	65.7	12.4	6.0
Basic research	18.4	15.6	53.8	12.2

Note: *Includes federally funded research and development centers (FFRDC)

Source: National Science Board – Science and Engineering Indicators, 2004

(e) US venture capital disbursements

Year	Total (billion $)	Seed (per cent)	Start up (per cent)	1st stage (per cent)	Later stages (per cent)
1995	7.635	4.72	12.63	23.43	59.22
1996	11.611	4.91	8.21	24.42	62.40
1997	15.080	5.24	3.62	23.22	67.92
1998	21.448	4.23	4.39	25.44	65.94
1999	54.785	2.23	3.82	22.14	71.81
2000	106.270	1.81	1.08	24.67	72.44
2001	40.825	1.05	0.93	22.36	75.66
2002	21.279	0.96	0.55	19.25	79.24

Source: Thomson Venture Economics, special tabulations, June 2003, reported in Science and Engineering Indicators – 2004

As detailed in Table 2.2 (d), while industry performed most of the developmental (89 per cent) and applied (66 per cent) research, universities perform most of the basic (54 per cent) research.[13]

A review of US human resource educational profiles[14] reveals that in 2002 more than 84 per cent of the eligible population had completed their high school education, and around 27 per cent held a bachelor's degree. In 2000, half of the US households owned computers; more than 40 per cent of those households had access to the internet; and according to a 2003 estimate the rate of broadband internet penetration in the US is 8 per 100 inhabitants – all of these percentages are quickly rising. According to 2001 estimates,[15] of the 140 million civilian labor force, there were 430 persons per 1000 workforce with recent bachelor's degrees in S&E and 89 persons per 1000 workforce with recent doctoral degrees in S&E. In 1999, of more than 13 million professional scientists and engineers (with 95 per cent holding bachelor's or higher S&E degrees) in the workforce almost 11 million (84 per cent) were employed – 70 per cent working in private industry, 17 per cent in universities and colleges, and 13 per cent for federal and state/local government. In 2002, there were 1.26 million S&E personnel involved in R&D activity in the US.

In terms of R&D performance, some indicators of basic research capabilities such as number of Nobel Laureates produced or citation analysis of scientific papers, and number of patents granted each year suggest that the research performance of the US system has been consistently strong.[16] During the Cold War years when resurgent economies of Western Europe and Japan developed significant technological capabilities that the US innovation system could not maintain high growth rates in real earnings nor could it enable national productivity growth to match these more successful industrial economies over longer periods of time. However, after the end of the Cold War, a renewed emphasis on commercialization and development of civilian technology caused the US economy to rebound by reversing some of these trends during the1990s.

In addition to the growing level of private sector R&D investment, the intensity and composition of industry R&D is changing. Since the early 1980s, the R&D intensity of US industry in aggregate has doubled. This is primarily due to the fact that R&D intensity of the knowledge-based high- tech sector has been increasing, and that most of the high-tech industries are growing faster than others, thus accounting for an increasing proportion of the US economy. Table 2.1 depicts the technology intensity of the US industrial base (2000) with its significant proportions of existing high-tech firms, number of employees, employee payroll, new births and high survival rates.[17] Among knowledge-based industries, there have been wide variations in R&D investment from industry to industry. Two key industries that are of special interest to this study – the electronics and information technology sector, and the pharmaceutical and biotechnology sector – have dramatically increased their share of industry R&D investments, and account for a large portion of the increase in overall industry R&D intensity. According to the National Science Foundation (NSF), the electronics and information technology sector's share of total industrial R&D investment grew from 32 per cent in 1981 to 44 per cent in 1995; in the same period, the drugs and medicine industry's share grew from 7 per cent to 16 per cent.[18] As a result, by the late 1990s, these two industries dominated industry-funded R&D growth in the US, accounting for more than half of all US industry R&D, and promising to drive future technology and innovation.[19] It should be noted that these two industries have benefited enormously from large federal R&D investments made on behalf of the government missions in defense, space and health. Shifts in industries' share of net sales also show the growing role of high-technology industries in the US, especially the role of the information technology and electronics sector.

In terms of the type of technologies attracting most of the venture capital in the US, communications, software and related services, semiconductors and electronics, biotechnology and medical, and computer hardware accounted for most of the disbursed dollars.[20] According to a NSF report, during 1990–2000, computer technology businesses received up to one third of all US venture capital funds, with software as the most favored technology area. And in spite of the recent downturn in the US venture capital industry (from $106.27 billion in 2000 to only $21.28 billion total disbursements in 2002), the top five industries that continue to receive the highest VC investment include software, information technology services, telecommunications, biotechnology, and networking equipment'.[21]

The venture capital financing may also be categorized by the stage at which money is more easily available. The trends show that very little venture capital is invested at the early seed level – for example, to an inventor, entrepreneur, or a small company trying to prove a concept. Over the 1995–2002 period, such seed money never accounted for more than 5.24 per cent of all US venture capital disbursements, and most often represented between 3–4 per cent of annual totals. A bigger portion of venture capital monies went to support product development and initial marketing – often referred to as start-up funds – but these investments still typically accounted for only 9–10 per cent of annual totals.[22]

A close examination of venture capital disbursements to companies over the last decade clearly shows (Table 2.2e) that most of the funds are directed to later stage investments, which captured more than two thirds of the disbursements. Capital for company expansion attracted by far the most investor interest.

2.1.2 The Federal System of Innovation

The US national R&D and innovation system consists of a wide variety of institutions and activities located in industry, universities, government, and non-profit organizations. Federal R&D has traditionally been performed by three sectors: the government itself, universities and industry. While the research relationship between the federal government and industry has historically been of great magnitude, it is only recently that the federal government has begun to participate broadly in cooperative science and technology programs. Historically, the research relationship between the federal government and industry has been defined on the basis of procuring or regulating technologies, research and products.

With respect to commercial technology, the federal government's relationship with the private sector was limited to research spin-offs, that is, technology first developed for government missions eventually making its way to the private sector for commercial application. From this base of huge government research and development expenditures arose America's global leadership in computer information/electronics, satellite communication, aerospace, and later in pharmaceuticals.[23]

There are, however, some notable and long-standing exceptions to this generalization. An early example is the Morrill Act (1862), which was the first major federally sponsored organized effort in the multi-sector cooperative programs. This effort, reinforced by the Hatch Act of 1887, and further strengthened by the Adams Act (1906) resulted in the Agricultural Extension Service (1914).[24] The Agriculture Extension Service had the express goal of aiding the well-being of individual farmers and was carried out through a country-based network of extension services. The precise design of today's cooperative models conceived to benefit small manufacturers differs, but the concept of a government-sponsored infrastructure to move technologies out of the 'farthest reaches of dusty farm roads' very much applies to contemporary industrial extension.[25]

The driving force for industrial extension in the federal government has been the National Institute of Standards and Technology (NIST). Founded in 1901 as the National Bureau of Standards, NIST has always filled a unique role in the federal government in standard setting, a powerful contributor to American economic growth and competitiveness. It has also historically collaborated with industry in cooperative research projects.

Between 1945 and 1950, as a result of a vigorous public debate on the institutional framework for science, the Office of Naval Research and the National Institute of Health were created. To capture the remaining general aspects of the institutional framework the National Science Foundation Act (1950) was promulgated. The NSF was created as an independent federal agency and was given the mandate to promote progress of science by advancing national health, prosperity and welfare, among other purposes.

The first significant shift by the federal government toward broad involvement in cooperative science and technology programs, well beyond the agricultural and standards related objectives, came in response to the US competitiveness challenges of the 1970s and 1980s. As a result, policies both at federal and state levels were revised to help establish a fuller and faster exploitation of government-funded R&D by US firms. This involved efforts to create partnerships between government-funded creators of technology,

Table 2.3 US cooperative technology programs by agency and type (1990s)

Program type → Fed Agency/Dept ↓	Technology development	Industrial problem solving	Technology financing	Start-up assistance	Teaming
Agriculture		√ ES	√ AARC, √ SBIR		
Commerce	√ ATP	MTC	√ ATP, √ SBIR	√ ATP	√ SLIPTI, √ MEP
Defense	√ MANTECH √ STTR	TRP	√ SBIR		NCHCP
Education			√ SBIR		
Energy	√ STTR	√ SBI	√ SBIR		
Health (HHS)	√ STTR		√ SBIR		
Labor		√ HRAP			√ NWAC
Transportation	√ IVHS		√ SBIR		
EPA			√ ETI, √ SBIR		
NASA	√ STTR	√ RTTC	√ AITP, √ SBIR		TUO
NSF	√ ERC, √ STTR,		√ SBIR		

	√ STC		
Cross-Agency	√ STTR	√ EPSCoR	√ FLC, √ NTTC

Note:
√ denotes existence of the program (for program abbreviations see list of Acronyms and Abbreviations)

Source: Berglund and Coburn (1995), p. 491, with author's verification of program existence, via web, August 2005

principally government laboratories and universities, and US industry to speed the development and commercialization of new technology. Since the 1980s, the US policy makers, both in the administration and in US Congress, have been continuously supportive of the ongoing numerous federal, state and local government initiatives covering all key sectors of the economy (Table 2.3). All these initiatives reflected a trend toward expanding the federal role beyond traditional funding of mission oriented research and development. They also reflected a trend toward facilitating technological advances to meet other critical national needs, including the need for economic growth that flows from the commercialization and use of technologies and techniques in the private sector.[26]

Several pieces of legislation that had a pervasive impact on cooperative technology development were promulgated in the 1980s. Until the passage of the Stevenson–Wydler Technology Innovation Act of 1980, which directed federal laboratories to help transfer their technologies to the private sector, technology transfer was not part of the mission requirements of most federal agencies. In the same year, the Bayh-Dole University and Small Business Patent Act was passed, which gave recipients of federal research grants better access to patents for federally funded innovations. Both acts initiated a period of profound change at federal laboratories and at universities throughout the United States.

In 1982, the Small Business Innovation Development Act established the Small Business Innovation Research (SBIR) program to increase R&D by small high-technology businesses. The act requires all federal agencies with large extramural research budgets to set aside a certain percentage of their R&D funding or grants (up to 2.5 per cent) for small firms.[27] Since its inception in 1982, the SBIR effort has become one of the most visible technology development programs at the federal level and has sponsored thousands of small technology-based businesses. In year 2000 alone 4361 awards worth $1070 million (2.5 per cent of all the extramural research) were disbursed in all the 50 states.[28]

The National Cooperative Research Act (1984), which was amended in 1993 by the National Cooperative Research and Production Act encouraged firms in collaborative research as well as production activities.

The Federal Technology Transfer Act of 1986 (FTTA) and the National Competitiveness Technology Transfer Act of 1989 expanded the scope of federal technology transfer, authorizing new Cooperative Research and Development Agreements (CRADAs) between federal laboratories and other parties, including private programs, universities, and government agencies at all levels. The intention of these CRADAs was to provide additional incentives for the transfer and commercialization of technology.

Out of these initiatives grew a dramatically new role for the renamed

National Institute of Standards and Technology. Under the Omnibus Trade and Competitiveness Act of 1988, the National Bureau of Standards became the National Institute of Standards and Technology (NIST). More substantively, the act established the Manufacturing Technology Centers (MTCs) program (for technology extension to small firms) and a program of research grants known as the Advanced Technology Program (ATP). While MTCs were created to facilitate the direct transfer of technologies from NIST programs to manufacturers, ATP provides seed funding to firms and consortia to accelerate the development and commercialization of high-risk technologies. The Manufacturing Extension Partnership (MEP) program created in 1993 combined MTCs with the State Technology Extension Program (STEP).

A second, parallel strain of the recent cooperative technology movements is the new emphasis on 'dual use' technologies and the quest for new civilian missions for the laboratories of the Department of Defense and the Department of Energy. This emphasis was motivated by the need to facilitate access by the US defense establishment to technologies in the commercial market, thereby reducing its own development and acquisition costs. It also sought to speed the application of defense technologies to commercial products as part of the role of bringing the military and commercial sectors closer together. The laboratories of the Departments of Defense and Energy are aggressively pursuing cooperative relationships with the private sector. To meet the common technology needs of civilian industry and national defense, US Congress passed the Defense Conversion, Reinvestment, and Transition Assistance Act of 1992. Under this Act, the Department of Defense, in particular, and the Advanced Research Projects Agency (ARPA) devised the Technology Reinvestment Project (TRP) to be administered by multiple agencies. The multi-agency TRP includes programs in three principal focus areas: technology development, technology deployment, and manufacturing education and training. It is coordinated by the Defense Technology Conversion Council, directed by ARPA, and operates largely through 'dual use' partnerships that involved private firms in any of a variety of participating federal agencies (ARPA, DOE, NIST, NSF, NASA, and the Department of Transportation).

These two strains reinforced each other powerfully. These cooperative technology programs are growing rapidly, and they are larger and more widespread today than ever before. Generically, these cooperative technology programs can be divided into five categories: technology development, industrial problem solving, technology financing, start-up assistance, and teaming. Although each category is distinct, there is overlap and coordination between them.

The federal government has undergone a considerable change in the past several years in its approach to the private sector. The broad awareness of and support for these activities in US Congress and their spread throughout the US federal R&D system ensure that they will continue well into the next US administrations. The role of the White House Office of Science and Technology Policy (OSTP) in cooperative technology programs has also taken on increased significance. The National Science and Technology Council has set government-wide priorities for several technology areas that include working cooperatively with the private sector.[29] The federal government represented by the White House Office of Science and Technology Policy (OSTP) and the National Governor's Association (NGA) have already signed a memorandum of understanding officially establishing the United States Innovation Partnership (USIP). The purpose of this agreement is to outline policies and procedures for a general working relationship to strengthen and nurture the USIP.[30]

2.1.3 The States' and Regional Systems of Innovation

Despite having a developed technology infrastructure and innovation support in most of the country, the infrastructural components remain concentrated in few states. Some states, often the most populous and wealthiest, and national regions, are technologically more active than others. The top ten states in terms of R&D expenditure are: California, New York, Michigan, Massachusetts, New Jersey, Texas, Illinois, Pennsylvania, Washington, and Maryland. With approximately half of the US population, they account for more than two-thirds of national R&D investments.[31]

For the purpose of this study we have expanded this list of top ten states to include a total of 25 states with significant levels of R&D expenditures and developed technology innovation infrastructure. Those 25 states account for three-quarters of the population and more than 80 per cent of the total Gross State Product (2000) of the country.[32] As shown in Table 2.4, these states, with 91.48 per cent R&D expenditure, disbursed 96.71 per cent of the venture capital in the country, and have 86.85 per cent high technology firms with 79.94 per cent of the high technology employment of the nation. Similarly, most of the science and technology degrees earned (80.34 per cent bachelor's, 84.97 per cent PhD), key technology clusters, the majority of science parks (78 per cent) and business incubation facilities (77.8 per cent) are found in various regions of these states.

Table 2.5 provides a summary of the past four decades of the state technology and enterprise development efforts (Plosila, 2004). As shown, beginning in the early 1960s, the US federal government encouraged a number of states to undertake technology development initiatives. In the first phase, under the State Technical Services (STS) program funded by the federal government, state advisory bodies were established in the 1960s, which lasted for about two decades. These advisory bodies helped policy makers address such issues as pollution, solid waste disposal and energy. Additionally, more progressive states such as Pennsylvania and Georgia supported industrial efforts based in universities, which evolved from the renowned agricultural extension model.[33] In 1977, the US Congress authorized the National Science Foundation to establish the State Science, Engineering, and Technology (SSET) program. The SSET program was intended to help the states develop and implement science and technology related strategic plans.

The second stage of state technology programs occurred during the 1980s. Technology innovation and development emerged as a major component of state and local economic development strategies, carried out largely without federal participation. The key element of the 1980s technology development programs was the establishment of university–industry technology centers.[34] A State of Minnesota study found that more than $550 million was expended on state science and technology efforts in the fiscal year 1988 alone.[35] While these dollar amounts are small in comparison to the federal government's support of research, they are significant when compared to past investments by states in these areas. Because many of these research programs are carried out at state-supported institutions of higher learning, the state government contribution to research is often much higher than the budgets contained in economic development functions.

During the third and most recent phase starting in the 1990s, the federal government began again to influence state program activities. Federal support for industrial extension and problem solving was reflected in several state program priorities. As federal cooperative technology programs grew in the early to mid-1990s, state efforts to secure federal funds by providing matching dollars and maintaining parallel programs increased dramatically. Also some states now emphasize that industrial problem-solving is a relatively low-risk approach compared to investing in a high-risk technology development strategy. Another explanation for the growth of the industrial extension model has been the increased emphasis among technology policy leaders on the importance of technology diffusion over technology development.

A Battelle study[36] reported that during 1990–94, the states spent a total of $303–385 million per annum to sponsor various cooperative technology programs including technology development, industrial problem-solving, technology financing, start-up assistance, and teaming through networks and information databases. This estimate, however, excluded related technology spending in traditional economic development, energy, transportation, health and support for higher education and training and so on. By the mid-1990s, states were spending about $2.7 billion of state guided funds on research and technology programs, which is 1.5 per cent of the total US R&D expenditure (SSTI, 1998) and more than the US National Science Foundation $2.5 billion budget (1998).

Table 2.4 Main R&D expenditure states with developed regional infrastructure for innovation

State	R&D Inv (million$) (2001)	VC Inv (million$) (2002)	High Tech Firms (2000)	High Tech Employment (2000)	S&T Parks	Business Incubators (2003)	S&E Degrees Bachelor/PhD (2002)	High Tech Clusters/Poles
California	50,959	9,467	60,799	1,397,776	8	123	22,403/3,232	Sacramento, LA, Other
Michigan	15,533	73	13,255	514,017	5	20	9,086/967	Detroit, Kalamazoo
Massachusetts	14,565	2,363	14,598	388,928	4	36	7,163/1,461	Boston Route 128
New York	14,422	803	27,507	513,472	11	76	15,180/ 2,124	NY Capital Region, NY City
Texas	12,722	1,284	28,410	703,206	8	43	12,941/1,462	Austin, San Antonio, Dallas
New Jersey	11,392	568	20,089	322,935	2	14	4,963/521	Middlesex-Somerset, Monmouth
Maryland	11,379	625	10,030	203,618	5	20	4,060/638	Baltimore
Pennsylvania	11,156	420	16,090	394,786	5	58	11,716/1,207	Philadelphia, Pittsburgh
Illinois	10,472	229	21,479	491,433	3	26	9,652/1,210	Evanston-Chicago Area
Washington	10,372	599	10,175	258,234	2	13	3,886/460	Seattle-Bellevue-Everett
Ohio	8,790	221	14,566	484,110	6	9	8,086/987	Cleveland-Lorain-Elyria
N. Carolina	5,825	547	10,887	268,284	4	34	6,300/697	Raleigh-Durh-Chapel Hill, Other
Florida	5,642	357	25,873	339,093	6	36	7,773/762	Orlando, Melbourne, W.Palm Bch
Virginia	5,554	409	14,015	348,426	5	34	5,993/603	New River Valley, Virginia, District of Columbia, and Maryland corridor
Oregon	5,447	159	5,693	108,254	2	9	2,503/233	Portland-Vancouver, Salem
Connecticut	5,311	219	6,356	166,788	2	7	1,759/353	New Haven-Bridgeport-Stamford, Connecticut
Minnesota	5,010	326	10,014	210,453	n.a.	26	3,998/403	Rochester, Minneapolis-St.Paul
Colorado	4,313	547	11,361	190,282	2	13	4,403/457	Denver, Bouldr-Longmt, Colo Spr
Indiana	4,235	52	7,049	302,599	3	22	5,620/589	Indianapolis

New Mexico	3,947	37	2,227	43,137	3	10	1,221/176	Albuquerque
Wisconsin	3,249	65	6,655	220,093	4	48	5,190/520	Madison, Milwaukee-Waukesha
Georgia	3,236	588	13,110	256,208	1	30	5,351/ 637	Atlanta
Arizona	3,048	191	7,493	166,678	2	4	3,571/ 417	Phoenix, Tucson
Tennessee	2,651	83	5,561	195,796	3	18	3,509/ 343	n.a.
Missouri	2,550	170	6,667	178,522	3	18	5,175/ 409	Kansas City, St. Louis
United States	253,355	21,096	425,992	10,050,578	136	961	213,467/ 24,558	
Above 25 States	91.48 %	96.71%	86.85%	79.94%	78%	77.8%	80.34%/ 84.97%	Per cent of US

Sources: State Science and Technology Indicators, Department of Commerce (2004), University Research Park Profiles, AURP (2003), De Vol Ross (1999)

In terms of their characteristics, state technology development efforts range from the dissemination of technological innovation to the establishment of large collaborative R&D centers. Common elements among these programs include (Mian and Plosila, 1996): (1) creation of a locus of responsibility for technology development in a state agency; (2) investment of state funds in research for long-term economic development; (3) a requirement for some form of leveraging of non-state resources; and (4) building of capabilities in one or more areas of science or technology.[37]

Table 2.5 State technology program evolution: summary overview (1960 and beyond)

Time period	Major players	Program emphasis
Phase one: 1960s & 1970s	• A number of states • Federal govt. through STS & SSET programs • Limited number of universities	State science and technology strategic planning, and science and technology advice from formal state advisory bodies
Phase two: 1980s	• Almost all states • A large number of universities • Private sector involvement	Emphasis on direct link to economic development. Cooperative technology development through university–industry research/ technology centers
Phase three: 1990s and beyond	• Almost all states • Several regional and local entities • A large number of universities • Federal govt. through NIST's cooperative technology programs	Shift from heavy emphasis on technology development to technology diffusion, support in industrial extension, and problem solving. More recently pro-active state/ regional strategies in leveraging technology for economic development, creating firm clusters and knowledge regions, with active higher education involvement

Note: For STS, SSET, NIST see list of Acronyms and Abbreviations

While state technology development programs have several common elements, they also differ in many respects: (1) the degree to which basic or applied research is emphasized; (2) the nature of the relationship and involvement of the private sector; (3) the manner in which state funds are distributed; (4) the degree to which single or multiple groups and organizations are involved in the operation of the program; and (5) the degree of accountability built into the program to access results over both the short-term and the long-term.[38]

Consequently, these programs vary in design, operation, funding support and structure. They may include diverse elements such as small business incubators, entrepreneurial development programs, venture capital, research parks, project matching grants, information dissemination, technology transfer mechanisms and centers of excellence. By the end of the 1980s, 44 states were involved in technology development support efforts and by the mid-1990s, 47 states had some type of technology programs.[39] By and large most state efforts were designed to intervene later in the innovation process than had traditionally characterized the federal science and technology policy, focused largely on basic research that preceded this. A few states such as Michigan and Pennsylvania began to provide services throughout the innovation process.

As to the general characteristics of the 1980s and 1990s-style technology innovation efforts, these programs offered states and localities a new paradigm of state and local economic development with several new emphases (Plosila, 2004): (1) risk: a new interest in risk-oriented programs including equity, seed, venture, and working capital investments; (2) credibility: a recognition that technology is an important component of assistance to traditional industries, including manufacturing; (3) higher education: an increased awareness that higher education institutions play a critical role in state and local economic development; (4) entrepreneurs: an increased interest in working with and involving entrepreneur-driven businesses, as contrasted to the past public agency focus on Fortune 500 firms during the 1960s and 1970s.

State universities have long been engaged in collaborative, pragmatic relationships with local industries. States have sought to promote economic development by establishing regional technology clusters, among the most notable being North Carolina's research triangle. During the 1990s, US states considerably increased their investment in technology policy, including university non-profit research centers, joined industry–university research partnerships, direct financing grants, incubators, and other programs in research and technology for economic development.[40]

Research centers supported by state governments have seen an evolution in thinking that has guided changes in design, structure and operation over the past several years. Initially, the state-funded research centers were primarily supported with direct lump-sum federal and state grants. Funds were made available for research on a block grant basis, with limited direction as to focus, content, or linkage to markets. In recent years, industrial involvement in centers has been through an industrial affiliate approach, but due to lack of deliverables and failure to adhere to timeframes for completion, such arrangements have increasingly been questioned by business regarding their true effectiveness.

New types of models and organizations are emerging in a few states to provide an alternative to the university research center or industrial affiliate arrangement. Consortia and networks are the terms generally used to describe these types of organizations. These newer approaches, more often found in Europe than in the US, are designed to serve small and medium-sized firms primarily; are market driven in design and implementation, with a focus on prototype design; provide a range of consultancy services to their industry members; and provide access to equipment and test beds.

These state and local technology development efforts represent 'novel experiments' and are important components in building state and local entrepreneurial economies. Small firms account for much of the innovation in the country and a disproportionate share of job growth. Large firms rely on technological innovations from small firms to maintain their competitive edge. State and local technology development efforts are expected to contribute at a sufficient scale and with significant impact, as they continue to refine their organizational and design characteristics – networks and consortia, use of local intermediaries, wholesale services and programs, investing and leveraging.

2.1.4 Summary and Conclusions

The past five decades have witnessed major shifts in the US institutional and policy framework for innovation. Following the end of World War II, there was an emphasis on building up the capabilities of national laboratories, universities, and corporate research centers and promoting research and development for government oriented missions such as defense, energy, space exploration, health and agriculture. In the 1960s, a focus on technology transfer emerged, with the aim to promote greater civilian spin-off from military-driven public research and development.

The late 1970s and 1980s debates over the US loss of competitiveness over economically resurgent Japan prompted initiatives in the 1980s to encourage the cooperative transfer of research and development from federal laboratories to industry, to establish university–industry research centers, and to promote industry consortia in electronics, machine tools and other industries. Through the Small Business Innovation Research (SBIR) Act of 1982 followed by the

Small Business Technology Transfer Program (STTR) ten years later, small business involvement in meeting federal R&D needs, while at the same time providing tangible support for technology commercialization into the civilian sector, was ensured.

Now, a third shift is underway. Prompted by increased global economic competition and the end of the Cold War, US technology policy is more attentive to explicit civilian commercialization goals. This cooperative technology movement includes a new emphasis on 'dual use' technologies and the quest for a new civilian role for government laboratories. Perhaps even more significantly, new networking-oriented implementation concepts are being pursued which, going beyond the traditional linear pipeline models of technology push and pull, are based on more complex and interactive perspectives on the technology development and diffusion process. In particular, newer patterns of industrial collaboration and commercialization are being promoted, to create industry consortia, university-industry linkages, and public-private partnerships. Examples of the new policy approaches at the federal level in the United States include the development of the advanced technology partnership ATP, the manufacturing extension partnership, the partnerships for a new generation of medical technologies, the United States innovation partnership, and the most recent nano-technology initiative.

These efforts to change the direction of the national innovations policy are not without opposition. Besides the difficulty of working with the public and US Congress, other factors at work have included the strong performance of the American economy in the mid to late 1990s and perceived need to boost defense spending in the post September 11, 2001 era, each in their own way keeping civilian technology policies away from the broader national policy agenda. Yet, these caveats notwithstanding, it seems that a new era in US technology policy has taken root, and the support of public–private collaboration and civilian technology goals will probably continue.

At the state and local levels, the 1980s and 1990s novel approaches and mechanisms for technology development through partnerships have increasingly been aimed at addressing the private sector market gaps. These initiatives have shown significant results by broadening their regional role in addressing technology, talent and capital issues which are likely to continue in the 2000s. While learning from their best practices, we are likely to see states further refining, modifying, and fully developing these and other initiatives. A rich menu of approaches and techniques may emerge that will facilitate the building of a better understanding of how best to sustain industry–university partnerships with states and federal government entities playing a catalytic role (Plosila, 2004).

In light of these facts, it may be concluded that the federal, state and local technology and enterprise development efforts will continue to remain pluralistic, undertaking a networking dimension. There are roles for the private sector, higher education institutions with enabling programs in matching grants, research centers, science parks, incubators, and seed and venture capital and so on. The upcoming decades are expected to further demonstrate the effectiveness of these experiments in the various states, regions, and communities of the nation as they mature and are further refined.

2.2 THE US CASE STUDIES

As detailed earlier, most US states have a well-developed technology infrastructure and related innovation support mechanisms, and technology-based economic development has become an integral part of state and regional agendas. Owing to their prior industrial history, resource endowments, and deliberate policy actions, half of the more densely populated states have developed advanced technology infrastructure with significant R&D expenditure. As shown in Table 2.4, more than three-quarters of the nation's innovation infrastructure is concentrated in 25 out of the 50 states, showing that they tend to be more active than other states in the pursuit of technology-based economic development (TBED). As listed in that table, some of these states have developed successful knowledge regions or TIPs, where local innovation patterns are having a pervasive influence on the relative rates of new firm creation and economic growth. Since the purpose of this research is to study those innovation poles that are making use of the modern incubation mechanisms, it was decided to focus on TIPs having relatively more established science parks and incubators – facilities with at least a decade-long track record of operating as key infrastructure elements in the emergence of their innovation poles. Table 2.6 provides a list of 15 such facilities (which roughly account for 10 per cent of all the science and technology parks in the US).[41] These facilities are generally believed to be representative of several planned incubation space models in the nation.

Following discussions of the selection criteria with several professional colleagues from AURP and other sources having expertise with science parks, a preliminary sample of seven facilities was selected out of the sampling frame of 15. These selected facilities have kept sustained innovation activities in their regions over more than a decade and are (a) equipped with both science parks and technology incubators (b) support a significant number of incubating firms forming sizable clusters (c) are models of the 1980s resulting from a variety of partnerships among various public, private and academic agents, (d) represent diverse rural, suburban and urban settings, and (e) include successful as well as not-so-successful cases (Figure 2.2).

This decision was communicated to the respective park managers to ascertain their interest and willingness to cooperate in a detailed case study. This led to the following four regional facilities agreeing to participate in the study: Rensselaer Technology Park in the New York Capital Region; Northwestern University Evanston Research Park in the Chicago Metropolitan Area; Virginia Tech Corporate Research Center in the New River Valley Region of Virginia; and the University Research Park in the Madison Metropolitan Area of Wisconsin. All of these facilities are considered comprehensive enough in providing most of the key elements of an enabling innovation infrastructure in the form of a science/ technology/ research park, technology business incubator, R&D centers and other technology and enterprise development support organizations generally affiliated with entrepreneurial research universities.

In each case, before we focus on the selected parks and their complementary facilities serving as technology innovation poles, we start with the surrounding metropolitan statistical area (or an equivalent region) that provides the ambient regional socio-economic context.

2.2.1 New York Capital Region

The New York Capital Region comprises the cities of Albany (New York state capital), Schenectady and Troy and is spread over an area of 3222 square miles (8345 square kilometers). The area has a population (2001) of 879 429 (with a 1.7 per cent increase during the 1990s) and a labor force of 402 100. It is one of the most densely populated regions in North America.

The area experienced a downturn in smokestack (iron and steel, automobile and so on) industries during the 1970s and 1980s triggering layoffs such as General Electric's loss of 4000 jobs in 1986 – some of GE's former employees established spin-off companies in the area. Paralleling the decline of industrial employment in the capital region has been a significant increase in employment in services, trade and small businesses in general. Thus with the transition of the US to a post-industrial society, the capital region has been evolving toward a 2-tier society, with an elite of PhDs working in research institutes and major universities in the area, and with unskilled workers, employed in low-value-added repetitive jobs and services, or unable to find stable employment. In the meantime, the blue-collar skilled working class, which has always been the link between the two classes and the backbone of local citizenry, is gradually but steadily disappearing – an ominous sign for the economy of New York's capital region. The next paragraph provides an overview of the region's key infrastructure elements that enable the region to support its main innovation pole described later.

Table 2.6 Established US science/research parks with regional firm clusters (rank ordered by year of establishment)

Park (year opened)/Region	Firms (number)	Employees (number)	Bldg Area Sq Ft (num bldg)	Incubator	Affiliated Universities
1. Stanford Research Park (1951) Sacramento, California	150	23 000	10 000 000 (162)	No	Stanford University
2. Cornell Business & Tech Park (1952) Ithaca, New York	90	1 600	567 000 (24)	No	Cornell University
3. Research Triangle Park (1959) Raleigh-Dur-Chap Hill, N. Carolina	131	38 000	19 000 000 (n.a.)	Yes	Three Universities
4. Cummings Research Park (1962) Huntsville, Alabama	220	22 500	8 500 000 (175)	Yes	U. Alabama-Huntsville
5. Univ City Science Center (1963) Philadelphia, Pennsylvania	150	7 000	1 714 000 (15)	Yes	Multiple (28) Universities
6. Univ of Utah Research Park (1970) Salt Lake City, Utah	42	6 000	2 420 000 (25)	No	University of Utah
7. Central Florida Research Park (1980) Orlando, Florida	80	5 500	1 300 000 (23)	Yes	Univ. of Central Florida
8. Yale Science Park (1981) New Haven, Connecticut	35	1 200	450 000 (8)	No	Yale University
9. Rensselaer Technology Park (1982) Capital Region, New York	**56**	**1 500**	**1 014 000 (18)**	**Yes**	**RPI University**
10. Research and Technology Park (1983) Seattle-Bellevue-Evert, Washington	12	70	50 000 (2)	Yes	Washington State U.

11. University Research Park (1984) Madison Metro Area, Wisconsin	**103**	**3 500**	**1 488 000 (24)**	**Yes**	**U. Wisconsin-Madison**
12. Corporate Research Center (1985) New River Valley, Virginia	**105**	**1 200**	**480 800 (9)**	**Yes**	**Virginia Tech U.**
13. Evanston Research Park (1985) Chicago Metro Area, Illinois	**80**	**1 500**	**400 000 (5)**	**Yes**	**Northwestern U.**
14. Univ Park MIT (1989) Boston Rt 128, Massachusetts	10	n.a.	1 425 000 (4)	No	MIT University
15. IC2 ATI Facilities (1989) Austin, Texas	38	n.a.	60 000 (1)	Yes	Univ. of Texas at Austin

Sources: AURP, 2003, updated through individual web pages, January 2005

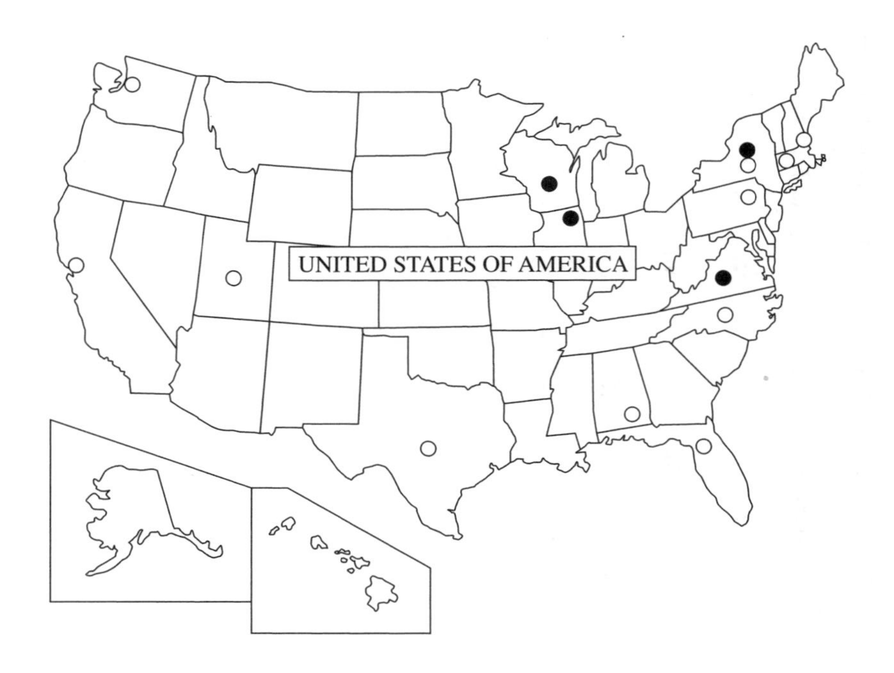

● Selected TIPs

NY Capital Region, New York;
New River Valley, Virginia;
Chicago Metro Area, Illinois;
Madison Metro Area, Wisconsin

○ Remaining TIPs

Seattle-Bellevue-Evert, Washington;
Sacramento, California;
Salt Lake City, Utah;
Austin, Texas;
Huntsville, Alabama;
Orlando, Central Florida, Florida;
Raleigh-Durham-Chapel Hill, North Carolina;
Philadelphia, Pennsylvania;
Ithaca, New York;
New Haven, Connecticut;
Boston Route 128, Massachusetts

Figure 2.2 US technology innovation poles targeted for survey

Regional Technology Infrastructure

There are more than 20 000 private (non-farm) business establishments in the NY Capital region with a 3.17 per cent state-wide increase in new establishment formation during the 1990s. Centrally located within the Northeast corridor, the region is ideally situated from a business standpoint. The region has a history of industrialization with a skilled, educated and disciplined workforce. There are 19 colleges and universities including

Rensselaer Polytechnic Institute (RPI), State University of New York (SUNY) at Albany and several technical colleges. SUNY Albany has 17 400 graduate and undergraduate students and RPI has 6878, with more than 200 PhD degrees awarded each year by the two institutions (2000). There are a number of private and state research institutions including General Electric (GE) R&D Laboratory; Albany Molecular Research Inc., a chemical–pharmaceutical research center; and Wardsworth Center, a comprehensive state health laboratory. These and other scientific and technical research institutions employ several thousand people (over 26 000, 35 per cent of total employment) with technical degrees including PhDs.[42] Given the area's harsh winters; the quality of life is considered moderate to high.

Located in Troy, RPI, with its strong history of involvement with technology and economic development realized this challenge. Starting in 1977 RPI pioneered the formation of multi-disciplinary research centers having 38 such centers by 2005. In 1980, under the leadership of its new president, George Low, RPI established an incubator program as a next logical step to promote technological entrepreneurship. This followed the establishment of a science park in 1982. Over the past two decades, these efforts have resulted in a host of additional institutional and programmatic elements in and around RPI, creating a comprehensive innovation milieu. The Troy area possesses a broad knowledge base with a developed infrastructure. Therefore, RPI emphasizes a diversity of technology sectors with strong information systems and manufacturing industry, though practically software and biotechnology are dominant.

In the Albany area, SUNY Albany too has research strength in several high-tech areas such as genomics and biomedical sciences, nanotechnology, microelectronics and advanced materials. Their cross-disciplinary R&D efforts aim to help develop the region's biotech industry and drive economic growth. To help commercialize these research results the university established two incubator programs in 1997. These programs offer student and faculty interactions with start-up companies and anchor tenants located in these facilities. There is a technology incubator named U-Start in Schenectady, which houses several start-ups. The Capital region's venture capital industry is fairly developed with $250 invested per worker; the NY State ranks sixth in the nation in VC availability.[43]

The region continues to offer a rich knowledge base and a highly educated manpower. Strong academic sector involvement with government and private sector support in the region has effectively leveraged this strength. The area's comprehensive technology development infrastructure with its active private sector and supportive government policies offers an ideal innovation milieu. Seed capital and venture capital funding is fairly developed. There are significant numbers of university-related entrepreneurs. Specifically, several of RPI's award-winning technology programs provide an innovation milieu that has become of world interest. The area's harsh winter climate often poses competitive challenges in capturing and retaining talent, yet its rich cultural and recreational activities and highly competitive schools tend to mediate this disadvantage.[44] The following paragraphs provide a detailed case study of the Rensselaer Technology Park as the main innovation pole of the region.

The Rensselaer Technology Park

The Rensselaer Technology Park (RTP) is a university-related science park, established in 1982. RTP is a division of the Rensselaer Polytechnic Institute (RPI), a private technological university, which is located in Troy, New York. The park is only five miles away from the RPI campus. This area is part of the New York State Capital Region with Albany as the State Capital. Situated along the Hudson River, RTP is conveniently accessible to the national interstate highway network, railroad system, and airports

Of the 1250 contiguous acres (506 hectares) of green land, which comprise RPT, only 250 acres (or 101 hectares) are developed for use. There are 23 buildings in the park with more than 870 000 square feet (81 000 square meters) space in use. The physical facilities are state-of-the-art with a road network and underground utility services including fiber-optic cabling. There are over 50 organizations residing in the park, most of whom are technology-based firms representing a wide array of technologies including computer science, biotechnology and electronics. There are over 2200 employees working for park-related firms. The park oversees its Incubator Program, which is located adjacent to the RPI campus, and where more than two dozen additional technology-based tenant firms receive start-up support. Table 2.7 delineates a summary of the survey data collected from the Rensselaer Technology Park (RTP) facility.

Park history and origins The Rensselaer Polytechnic Institute (RPI) has had a history of strong involvement with technology and economic development. Established in 1824, RPI is the oldest non-military engineering college in the Western hemisphere, founded in an area considered the birthplace of the Industrial Revolution in North America. From its very origin, its founder Stephen Van Rensselaer sought to provide an educational environment where persons could 'apply science to the common purpose of life'.[45] Today, RPI has a student body of about 6000 undergraduate and graduate students who are enrolled in one of five academic schools: Architecture, Engineering, Humanities and Social Sciences, Management and Science. As a technologically oriented institution of higher education, RPI pioneered in 1977 the formation of multi-disciplinary research centers, bringing together experts in the field of engineering, physics, chemistry, computer science, and

Table 2.7 Rensselaer Technology Park: summary data

Sponsor(s)	Rensselaer Polytechnic Institute
Year opened/settings	1982/Suburban
Size: Land/buildings	1250 acres (506 hectares) with 250 acres (101 hectares) in use/23 buildings with 870 000 sq. ft. (81 000 sq. m.) space
Start-up support (Incubator)	Separately located 'Incubator Program' with lease area of 70 000 sq. ft. (6503 sq. m.)
Affiliated organizations	Venture Affiliate Program, University Economic and Technology Development, Office of Technology Commercialization, Entrepreneurship Center
Resident organizations including TBFs	Park: around 50, Incubator: 43
Tenant employees	About 2200
Specialty/focus	Technology transfer and commercialization of a broad range of university technology
Mission	To attract a broad diversity of technologies in areas reflective of strengths of RPI's interaction for economic development
Organization and governance	The Park is a division of RPI. Governance is by a standing committee of University Board of Trustees. The incubator reports to RTP Director, who reports to the University President.
Financing, capitalization, revenue	RPI endowment, state support
Marketing – targeted technologies and type of entrepreneurs	All high-tech areas are targeted; almost half of the entrepreneurs are from faculty, students and alumni
Tenant services	Business and technology development services, seed capital, VC access, library and databases, university facilities, R&D, assembly and light manufacturing allowed in the park
Technology-based firms surveyed	Out of 74 resident organizations, 24 TBFs were surveyed with 7 respondents (29% response rate)

Source: RTP Survey

so on to study such critical problems as manufacturing productivity, and to help define the technologies of the future with respect to integrated electronics and interactive computer graphics. As stated earlier, 38 such centers are operating within the university system which focus on a technology portfolio consisting of chemical processes, mechanical processes, mechanical testing and systems, electronics, microelectronics, environmental sciences, life sciences, medical devices and new materials.

At the same time, RPI boldly embarked on a new direction for higher education by creating opportunities for the seeds of technological innovation formed in its research laboratories and interdisciplinary research centers, to develop and grow into viable technology-based enterprises. Building on its technological reputation coupled with the vision of its President George Low, a number of new programs were launched in the late 1970s and early 1980s. First, RPI undertook a feasibility study for the development of a technology park in 1979. In 1981, after selecting a 1200-acre parcel of land owned by the university in the nearby North Greenbush area, the University Board of Trustees authorized $3 million for the first phase of infrastructure and associated management costs to develop the park.

The Incubator Program was founded in 1980 and is located in three campus buildings within a radius of five miles from the park. With a lease area of 70 000 square feet (6503 sq. m.) the Incubator strives to provide a nurturing environment to a select number of technology-based start-ups by offering a wide array of shared and professional services and networking support. The program connects entrepreneurs to university and community resources; a facility accesses to a broad-based network of businesses, financial and technical experts, and provides education and training programs focused on tenant entrepreneurs.

The RPI also understood the need for synergy between the university and the community by recognizing that the resources of the university could have a positive impact on the economy. Over the past two decades, these efforts have resulted in a host of additional institutional and programmatic elements in and around RPI, creating a comprehensive innovation milieu for technology-based entrepreneurial economic development of the region. Besides the Technology Park and the Incubator Program, the other core elements of the innovation milieu that were added include: the Center for Entrepreneurship of New Technological Ventures, the Rensselaer Technological Entrepreneurship Council (RenTEC), and the Capital Region Technology Development Council.

The RPI's Center for Entrepreneurship and New Technological Ventures was established in 1988 to train future entrepreneurs. The program is housed in the School of Management and Technology, now offering more than 18 courses that enable at least 500 students per year to learn the fundamental principles and operations of managing entrepreneurial ventures.

This followed a new initiative named Rensselaer Technological Entrepreneurship Council (RenTEC) in which all of these innovation programs were united into one team in order to focus the combined strength/ expertise on technology commercialization. This aims to draw upon the entrepreneurship infrastructure in order to aggressively pursue university-developed technology through licensing and new venture development.

Early in the decade of the 1980s, RPI President Low, together with businesses, political and education leaders of the New York State Capital Region, formed the Capital Region Technology Development Council with the express purpose of retaining and nurturing high technology businesses in the Capital Region. This regional initiative led to the formation of Technology Development Organizations (TDOs).

Each of the above programs has played a major role in the quality of the university as well as a new role for RPI as a contributor to the technological entrepreneurship focus for growth of the regional economy. In fact, over these years, a new university paradigm has emerged from the Rensselaer experience in which universities have recognized that they have a role to play in economic development and that this role does not conflict with the mission of the university. These programs, particularly the Technology Park and the Incubator, were pioneers in exploring new and interesting economic development initiatives, which, in the past two decades have evolved from a novelty to an industry flourishing throughout the United States, North America and the rest of the world.

Tenant firm characteristics At the time of our survey,[46] there were more than 72 tenants in the Rensselaer Technology Park – over 50 tenants in the park and another 22 in the Incubator. These tenants represent a wide diversity of technologies from electronics to physics research, from biotechnology to software. There are some service providers such as educational material, business and technology consulting, as well as legal advice.

According to the survey, 28 per cent of the tenants (21 firms) are software and information systems technology related, and 39 per cent (28 firms) are service related, which is a typical tenant profile of a national research park having a technology incubator. Other technology areas with significant representation are biotechnology, environment and ecology, new materials, instruments and control. For the purpose of this study a population of 24 firms (10 from the park and 14 from the incubator), which fall in the category of technology-based firms were targeted. However, only seven firms (29 per cent return) participated in the survey.

The research shows that almost three-quarters of the surveyed firms have been established in the 1990s with all but one of these having a life of five years or less (57 per cent). Over the past five-year period, about 11 product (both goods and service), four process, and two management-related innovations were reported. About 44 per cent of the respondents said that market pull was the key driving force of these innovations, while 25 per cent believed technology-push was the key driving force, and another 31 per cent said both market pull as well as technology push were the contributing factors. In terms of sources of innovation, half of the respondents said that internal sources were responsible for the success of these innovations. A majority (62 per cent) of these innovations made use of patents, and to a lesser extent they used design, and trademarks respectively. Most of these firms are privately held national firms, and a small number are incorporated, or are subsidiaries of large firms. The market coverage of their products/services is both national and overseas as reported by a majority of the tenant firms surveyed. About half of the firms that reported their past five-year sales showed growth while the other half did not report their sales figures due to privacy reasons. In terms of R&D workforce concentration, when put together, the surveys show two and a half times as many personnel engaged in R&D than in other areas of business (sales, marketing, management and so on).

This analysis shows the typical profile of a technology-based firm being incubated in the RTP facility. While almost half of the surveyed firms are housed in the Incubator, the remaining half belong to the park. Some of the more prominent corporate tenants in the park include: MapInfo Corporation, a successful desktop mapping software developer operating globally, and Coromed Inc., another global firm providing clinical research services to the medical/pharmaceutical industry.

Performance and regional impact Some facts about RTP would be helpful in determining its performance and resulting impact on the region. This impact is measured along three dimensions: (a) degree of realization of its mission (b) financial self-reliance (c) promotion of tenant firms.[47]

It is obvious that the companies are attracted to the park for two reasons. First, they are involved in some way with science or technology – they use it, create it, or service it. Second, they wish to enhance a relationship with RPI for additional benefit of the entrée and connection to the university. Well over 90 per cent of the park firms have some relationship with Rensselaer. These results seem to be in line with RTP mission.

A vast majority of the park and Incubator companies are involved with not only RPI, but other universities in the area as well. Several of the park's 50-plus current tenants got their start at the Incubator. The park, with its 50 tenant firms, employs about 2200 workers, with $55 million company payrolls, and generates $1.7 million annually in property taxes.[48]

The Incubator Program has a mandate from RPI to operate the program as a cost center on a break-even basis. Therefore, unlike many other incubators in the nation, RPI Incubator has remained financially self-sustaining over the past several years.

Over 150 companies have been served since 1980 and, according to the management, more than 80 per cent of the participating companies in the Incubator have survived. Additionally, most graduating companies have stayed in the New York Capital Region. Therefore, the impact on the region with the reported almost 90 per cent incubated companies remaining in the region is substantial. The Incubator tenant firms provide 230 jobs and average $124 million (2000) in aggregate annual sales. Owing to the popularity of the program, the occupancy rate is almost 95 per cent. Annual sales of incubator 'graduates' exceed $200 million (2000) according to one estimate. According to another estimate, around two-thirds of RTP firms have originated either from research at RPI or have been started by Rensselaer Alumni – over 10 per cent of the park's employee base is made up of alumni, and over the years RTP firms have employed hundreds of RPI students. As a result of this achievement, the Incubator Program has been a recipient of the National Business Incubation Association's 1995 Incubator of the Year Award. Likewise, The Association of Research Parks honored RTP with its award for outstanding research park achievement in 2000.

Conclusion The New York Capital Region, with its educated and disciplined workforce and developed innovation infrastructure, is repositioning itself as a region that offers a knowledge-laden milieu. The region's main science park, RTP, has established itself as a successful TIP offering a more focused incubation space in the region. There are respectable sized firm clusters in the software and biotechnology areas. The continued leadership role of RPI in the form of its incubation mechanisms, and increasing participation by SUNY-Albany and other institutions of higher education in the science and technology-based economic development, are healthy signs for the region. The active private industry involvement and supportive government policies have also been encouraging. Since the region possesses a broad knowledge base with a strong information system and manufacturing industry, and fairly developed seed and venture funding programs, it warrants a continued emphasis on diversity of technology sectors. This comprehensive technology development infrastructure, with active private sector and supportive government policies, offers a well-developed innovation milieu.

For sustainability and future growth, the current strong academic involvement in the region has to be maintained and effectively leveraged in harnessing its traditionally rich technology base. Specifically, the ongoing efforts in implementing best practices at RPI's award-winning technology incubation mechanisms such as RTP, and the newly emerging nanotechnology-related innovation centers at SUNY-Albany[49] offer a bright future for a vibrant TIP that has already become a model for emulation. The area's high cost of doing business because of higher taxes, and its harsh winter climate often pose competitiveness challenges in capturing and retaining talent; however, its negative effect can be mediated by continuously improving government policies and the provision of quality of life amenities.

2.2.2 Chicago Primary Metropolitan Area

The Chicago Primary Metropolitan Statistical Area (PMSA) includes the counties of Cook, DuPage, Kane, Lake, McHenry & Will, DeKalb, Grundy and Kendall. It is the third largest metropolitan area in the US with a population of 8 million people and a land area of 5085 square miles (13 170 square kilometers).

Of the 500 largest US corporations listed in Fortune Magazine, 46 are located in the Chicago area and most others have regional branch offices located in Chicago. The city is one of the world's largest centers of commodity trading activity and is the home of the country's largest exchanges. Additionally, Chicago and its neighbor, Gary, Indiana are locations of several large steel mills and their associated manufacturing industry groups. The adjoining city of Evanston has a well-developed medical service sector. While the City of Chicago has registered a decline in its population over the past few decades, the overall Chicago PMSA population has shown a modest increase of 2 to 5 per cent over the past two decades, owing primarily to rapid growth in the neighboring counties. The Chicago PSMA with its $25 865 per capita income ranks among the top 12 in the US.[50] It is centrally located and has a stable, exceptionally diversified, and healthy economic base. The area's non-farm employment is 3 969 700, of which 80 per cent is service-related including government, and which increased by 20 per cent during 1970–90. The next paragraph provides an overview of Chicago PSMA's key infrastructure elements that enable the region to support its main innovation pole described later.

Regional technology infrastructure

The Chicago PSMA is rich in higher education resources with a large number of skilled and highly educated workers. Public school enrollment was around 1.2 million during the 1990s. There are several universities and colleges, including the University of Chicago, Northwestern University, and the University of Illinois at Chicago. These three universities have a total of over 30 000 undergraduate and more than 18 000 graduate students, and they award several hundred PhD degrees each year. The University of Chicago operates Argonne National Laboratory for the US Department of Energy. The two organizations together comprise the largest research enterprise in the Midwest, conducting more than $700 million in sponsored research (2000). They have licensed more than 150 technologies to the private sector, and entrepreneurs have created more than two dozen companies that are based on licensed technologies developed at the University and Argonne Labs. The technology transfer arm of these two organizations is called ARCH (derived from Argonne-Chicago).

In the 1980s, faced with the region's stagnant economy and higher taxes, the need was recognized to leverage university technology to help diversify the area's service economy in order to create jobs. The adjacent city of Evanston, located in the northern part of Chicago Metropolitan Area, needing an enhanced tax base, teamed up with Northwestern University to develop a technology infrastructure – Incubator (1982) and Science Park (1985). This area offers an extensive R&D-intensive knowledge base, and a highly trained and educated workforce with an entrepreneurial climate. Proximity to the university as a source of knowledge and new technology is the primary attraction of the park location. Northwestern has long been a leader in the creation of interdisciplinary research centers responsive to industry needs. Three of these centers, including an ARCH facility, are located in the research park (discussed below). The US Department of Energy-funded Basic Industrial Research Laboratory (BIRL) serves as an anchor tenant in the park. With the area's highly networked technology infrastructure and market-driven mixed use, Evanston Park and other area incubator facilities offer a sustainable innovation milieu. The venture and seed capital resources are fairly well developed with Evanston Inventure and other Chicago area VC firms.

The Evanston area has a high-income, educated population and a strong service sector which replaced smokestack industries. University technology transfer and commercialization in software, biotech and materials dominates. The area's successful incubation activity has generated a high-value-added diversified economic base with several newly incubated tech-based firms providing a potential for future growth in this knowledge-laden environment. Although the climate is harsh during winter, in all other measures the area has a high quality of life with amenities. The evolving nature of the academic–city government partnership, especially in Evanston, continues to pose challenges. The following paragraphs provide a detailed case study of the Northwestern University/Evanston Research Park as a main innovation pole of the region.

Northwestern University Evanston Park

The Northwestern University/Evanston Research Park was established in 1985. The University and the City of Evanston collaborated in the development of this semi-urban park space, which is situated on a 22-acre (9 hectares) triangular plot adjacent to the Evanston central business district and the University's administrative complex. With all but three of the 18 sites developed, the park has well over 400 000 square feet (37 160 square meters) of technology space that has been totally leased or sold. A recently completed commercial development portion added an additional 750 000 square feet (69 675 square meters). More than 1000 people are currently employed in the technology firms housed in the park, and the vast majority of these jobs are new to Evanston, a small city of 75 000. The Research Park (including its incubator)[51] has 62 tenants with a concentration of software, information systems, medical services and biotechnology firms.

Table 2.8 provides the summary of the research data about the Northwestern University Evanston Research Park (NUERP) facility. The table shows a rich array of support mechanisms in and around the Northwestern campus, providing a rich milieu for commercializing research results, the primary focus of the park.

Evolved as a successful mixed-use semi-urban facility, the park is a salient example of university, industry and government partnership for technology and economic development in the nation. The facility has drawn from the university's research strengths in science, engineering, medicine and business along with benefiting from its proximity to the vibrant Chicago metropolitan area. As the following paragraphs will show, the local and regional economic impact of this well organized and professionally managed facility has been pervasive.

Park history and origins Northwestern University and its host, the City of Evanston first discussed the concept for an urban research park in the early 1980s. This University–City joint project envisages a mutually rewarding partnership in developing a longer-term economic development strategy with the following perspectives.

The University saw the park as a vehicle to develop linkages with the area's burgeoning business community to promote technology transfer and gain opportunities for student employment and training, while at the same time earning a reputation as a partner in the area's economic development. The City's motive for taking on this offer was to develop a parcel of unproductive land adjacent to the University and downtown Evanston to revitalize the City and gain new jobs and property tax rolls.

Northwestern's initial commitments to the park included: (a) university-owned land in the development area (approximately six of the 22 acres); (b) construction and operation of a major anchor facility, the Basic Industry Research Laboratory (BIRL), with initial funding ($26 million) from the US Department of Energy; (c) space in a university-owned building to house a small business incubator; and (d) a $4 million line of credit to the City for property acquisition and infrastructure improvements. In addition, the

Table 2.8 Northwestern University/Evanston Research Park: summary data

Sponsor(s)	Northwestern University and the City of Evanston partnership
Year opened/settings	1985/Semi-suburban
Size: Land/buildings	22 acres (9 hectares)/15 buildings, 400 000 sq. ft. (37 160 sq. m.) technology space; 750 000 sq. ft. (69 675 sq. m.) commercial space
Start-up support (Incubator)	Separately located 'Technology Innovation Center' with lease area of 54 000 sq. ft. (40 969 sq. m.)
Affiliated organizations	Evanston Inventure, Basic Industry Research Lab, Basic Industrial Research Institute, Center for Biotechnology, Small Business Development Center, Institute for Learning Sciences, Institute for International Entrepreneurship
Resident organizations including TBFs	Park: 25, Incubator: 37
Tenant employees	About 1000
Specialty/focus	Technology transfer and commercialization of university technology in software, biotechnology, and materials
Mission	To effectively commercialize university technology and provide economic base to the city of Evanston
Organization and governance	The Park is a for-profit corporation equally owned by the University and the City. The incubator is a not-for-profit corporation reporting to the Park
Financing, capitalization, revenue	Northwestern University, City of Evanston, the State of Illinois. University share of $22 million
Marketing – targeted technologies and type of entrepreneurs	Predominantly information technology especially software, biotechnology/medical; small number of faculty and alumni entrepreneurs
Tenant services	Business and technology development services, seed capital, VC access, library and databases, most university facilities, R&D allowed, assembly and light manufacturing not allowed in the park
Technology-based firms surveyed	Out of 94 resident organizations, 31 TBFs were surveyed with 9 respondents (29% response rate)

Source: NU/Evanston Park Survey

University allocated substantial staff resources to the project in the areas of planning, marketing, technology transfer, and small business development.

The City's commitments included: (a) City-owned land within the development area, as well as a commitment to acquire all remaining parcels within the bound of the park; (b) new infrastructure, including sewers, water, street creation and reconstruction, lighting and parking; (c) creation of a tax increment financing district to generate funds for necessary land acquisition and capital improvements.

In 1986, Northwestern University and the City of Evanston formed the Research Park Inc., a for-profit real-estate entity, as equal partners. A private real estate company was assigned the responsibility to develop the 22-acre land into park buildings. The first multi-tenant park building with 43 000 square feet (39 947 square meters) leasable space was commissioned in 1988. The university had already made available one of its older buildings with 11 000 square feet (1022 square meters) rentable space to establish a business incubator.

While these developments were underway, sensing the opportunity, a group of the City's largest employers established a local financing entity named Inventure, which raised $1 million as seed fund. To establish its presence in the park, Inventure also obtained State of Illinois funding to establish a Small Business Development Center (SBDC). Both the seed capital fund and the SBDC operated within the park. In 1987, the university secured a $25 million federal government grant to build a 130 000 square foot (12 077 square meters) Basic Industry Research Laboratory (BIRL), which became the technology anchor in the park. The overall responsibility for BIRL was not only to act as an anchor tenant but also to serve as a bridge between the university researchers and the area's technology-based businesses.

In 1990, a second multi-tenant building with 56 000 square feet (5202 square meters) was added to the park and a rehabilitated warehouse building with 55 000 square feet (5110 square meters) was provided to the growing university technology incubator. The incubator also received a State of Illinois grant of $300 000 to fund its expansion. At this initial stage the incubator received an operational subsidy of approximately $75 000 per year from the park. This followed an extension of the university's Ethernet network into the park and its incubator, providing fiber-optic connectivity to tenant firms.

By the mid-1990s the City and the university were renewing their partnership commitment to a four-year extension as partners and they created a new private corporation in the form of Northwestern University/Evanston Research Park, Inc. (NUERP, Inc). They also noted that the park's progress was stymied owing to a lack of additional space for expanding businesses, which started moving out of the area. This prompted a change in strategy for land development. It may be noted here that the initial park plan had envisioned a mixed use project – including commercial and residential development alongside the technology space. To address this challenge, the partners agreed on a mixed use development strategy to support the development of its remaining sites.

By the end of 1990s the park contained approximately 400 000 square feet (37 160 square meters) of technology space, housing some 60-plus technology-based companies and over 1000 employees. At the time of the survey 15 out of 18 total sites were already built. The incubator lost some technology-based tenants due to the fact that the rental costs became unaffordable for them when the subsidies were withdrawn. Efforts are underway to convert BIRL building into a technology incubator focusing more on university technology commercialization objectives. The downsized incubator houses around 35 companies in its two buildings.

After the withdrawal of the City from this partnership at the end of the 1990s, the core of the park with its technology activities was concentrated in the north end, with six multi-tenant buildings in a two block area next to the university. The south end of the park has a modern hotel, restaurants, a cinema, condominiums and offices. In its present form the park may be considered as a mixed-use new urban model of a knowledge pole.

Tenant firm characteristics For the purpose of this study, a population of 31 firms falling into the category of technology-based firms was surveyed.[52] However, only nine firms (29 per cent return) participated. The fact that out of the 62 current tenants in both the Park and the Incubator, one third (20 firms) are software and information systems technology-related, and another quarter (15 firms) are service providers, conforms to the national trend in the research park tenant statistics.[53] Other technology areas with significant representation are health and pharmaceuticals, biotechnology, optics, instruments, and engineering design.

The firm survey shows that they were almost all established in the 1990s with more than half (55 per cent) within the past five years. Over the past five-year period, about 23 product (both goods and service), 13 process, and four management-related innovations were reported. About half of the respondents said that market pull was the key driving force of these innovations, while another quarter each expressed the role played by technology push or both (equally by market as well as technology push) as the contributing factors. In terms of sources of innovation a majority of the respondents said that internal sources were responsible for the success of these innovations. A majority (56 per cent) of these innovations reported making use of patents, and to a lesser extent they used design, trade secrets, and trademarks. Most of these firms are national corporations, with a smaller number of single owned firms or partnerships; only two of the responding firms had a foreign share in their ownerships. The market coverage of their products/services was reported to be both national and well as overseas in most of the cases, with the exception of only one case with a national coverage only. All of the technology-based firms studied reported positive trends in their sales figures over the past five years, proving them to be successful operations.

After leaving out one multinational biotech firm that reported a higher percentage of a non-R&D work force, all others when put together show an average of twice as many personnel engaged in R&D than in other areas of business (sales, marketing, management and so on).

This survey shows a typical profile of a technology-based firm being successfully incubated in a technology incubator/science park. While two thirds of the surveyed firms are housed in the incubator, the remaining one third belong to the park. One of the prominent corporate tenants in the park include the Midwest office of IDEO, the nation's leading design and engineering firm, specializing in technology-based products – it has offices in Palo Alto, Paris and Tokyo. The IDEA started this office in the incubator with one person and today it employs 33. The other prominent tenant firms include: Center for MR (Magnetic Resonance) Research, a comprehensive magnetic resonance facility owned by the Evanston Hospital Corporation; Celsis International, a UK-based rapid microbial diagnostic system developer for bio-medical manufacturers for use primarily in the food and cosmetics industries; and Fujisawa Research Institute of America, a research and analysis facility specializing in drugs that affect the immune system. Companies 'graduate' from the incubator when they reach the ability to function on their own. Most incubator companies have relocated either into other park buildings or outside the park into Evanston.

Performance and regional impacts A review of the overall performance of Northwestern University/Evanston Research Park shows a number of successes.[54] The City–University partnership did fulfill some of the key objectives: for the university in developing relationships with the area's technology businesses, and for the City in opening new businesses and jobs along with an expanded tax base as a result of developing its 22 acres of unused land. Over the past more than one and a half decades, the park has helped develop a new economic base for the City of Evanston, adding its technology businesses to health care, business services, institutional, retail and entertainment uses as the local businesses and employers of the future. On the emerging south side of the city, the theaters, restaurants and retail stores are successful and bustling. The combination of technology-based businesses along with urban development seems to be the new mixed-use direction providing sustainability and growth for the park.

Since its inception, the park has accommodated up to a maximum of 90 companies. The incubator, which is located next to the park in separate buildings, has generated over 165 companies and today more than half of the park tenants moved from the incubator after graduation. The park has dramatically impacted the economic base of Evanston as well as the city's image. During the past 17 years the university has expended more than $22 million for economic development and technology transfer in the park.[55] The commitment in turn has helped Northwestern generate substantial new public sector support for its applied and basic research, funding for research facilities on the campus, and has stimulated increased interaction with the private sector. Operationally, the incubator is largely self-supporting, with 87 per cent of the expenses covered from rental income and the remaining 13 per cent coming from the park's financial support. According to an estimate, over the first 15 years of its operation the incubator tenants and graduates, which include some high growth companies, created over 1500 full-time jobs with gross revenues of $120 million per year. At the time of this survey, however, the incubator tenants and its graduate firms employed 665 full-time equivalent employees who paid around $622 000 in taxes. One of the goals of the Research Park development has been to leverage taxes for the City of Evanston. Today, the revenues exceed $2 million per year, and the total revenues collected since 1987 equal more than $12 million. These dollars have been used to repay the City's redevelopment expenses for the Research Park.[56]

Northwestern University students and faculty have enjoyed collaborations with Research Park companies through internship programs, company creation and research partnerships. Today, six faculty members and 17 students and/or alumni have founded companies in the park and, since 1994, 73 undergraduates have worked in companies through a State of Illinois sponsored internship program which provides matching funds to companies that hire Illinois resident undergrads at Northwestern. Most importantly, research park companies have provided active learning laboratories to augment the curriculum in various degree programs. Over the past ten years, the park has generated over $2 million in minority and women business contracts, exceeding hiring goals. Out of this success came the recent creation of a minority and women's business consortium made up of Northwestern University, the local hospitals, school districts, Research Park, and the City of Evanston – all joining to facilitate contracting, hiring and training. These combined resources have led to the creation of many jobs. At the time of this study[57] there were 38 undergraduate, 47 graduate students and 12 faculty members working for companies within the Research Park – almost all companies in the park have hired or are using Northwestern students as part of their manpower requirements. Proximity to the University is the primary attraction of a Research Park location. Tenant firms enjoy ready access to a multitude of university resources including faculty and graduate students, as well as access to laboratory and library facilities.

Beyond the bounds of Evanston and the Chicago metropolitan area, the Research Park has drawn the attention of foreign delegations from around the globe. Overseas and national visitors interested in research and technology transfer, economic development, university-industry relations, and business incubation travel to learn from park staff what works and what may not work in the creation and development of research parks.

In spite of the park's inability to develop new construction from 1992 to 1996, its incubator program continued to grow. Companies 'graduate' from the incubator when they reach the ability to function on their own. Some notable graduate firms are: Illinois Superconductor and Peapod, both of which are traded on NASDAQ. Other notable companies are: Plextel; Metabolic Technologies; Recor; Real Fans; Perceptual Robotics; Student Advantage; U-Access; QuesTek; and MediFacts. Other than Peapod and Illinois Superconductor, the other 12 incubator graduates have relocated either into other park buildings or outside the park within the City of Evanston.

To recognize this success, the National Business Incubation Association gave the 1997 Incubator of the Year award to its incubator program.

Conclusion The Chicago Metropolitan Area, with its stable and well diversified economic base has emerged as one of the more promising knowledge regions offering a sophisticated innovation infrastructure. In the Northwestern Evanston Research Park, university technology transfer and commercialization in software, biotech and materials has taken root, resulting in several success stories of NTBFs generating a high-value-added diversified economic base for the region. The region's highly networked technology infrastructure and market-driven mixed-use park and other area incubator facilities offer a sustainable innovation milieu. The venture and seed capital facility is fairly advanced.

Currently housing over 60 TBFs with about 1000 employees, the park support a steady stream of new start-ups in targeted technologies and has developed a solid entrepreneurial support infrastructure. NU/ERP has become a rallying point, a public symbol for the importance of the partnership between university and the host City for the purpose of developing a knowledge-based economy. However, the evolving nature of an academic–city government partnership is not without challenges as the park emerges as a modern mixed-use urban facility.

In summary, Northwestern University Evanston Research Park, and particularly its incubator component, is considered as one of the nation's more established technology incubation programs, offering a well-developed infrastructure to nurture new technology-based firms. Since its inception the park has emphasized its role as the promoter of technology-based high-value-added businesses in this economically challenged region of the industrial 'rust belt'. As a novel experiment of university–industry–government partnership for technology development during its first 15 years of existence, the park's focus has recently shifted toward more of a mixed-use facility in order to attain financial sustainability. The park has achieved several milestones in its more than one and a half decades of existence, and more recently has been relying more heavily on its incubator component in search of its coveted science park identity.

2.2.3 Virginia's New River Valley Region

The New River Valley (NRV) is predominantly a rural region of Southwestern Virginia, which includes counties of Montgomery, Floyd, Giles, Pulaski, and City of Redford. The total land area is 1468 square miles (3802 square kilometers) and the total population of the NRV region is 165 146 (2000).[58] Historically, this rural area lacked any major private technology-oriented businesses and there was no viable industrial base. Surrounded by natural beauty, the area is now developing by maintaining a growing economy, strong technically oriented higher education and an increasingly attractive business environment. The area's burgeoning population is thanks to a favorable climate and central location; the presence of Virginia Tech, a major research university with an R&D-intensive environment prompted the plan for economic diversification.

The region is rich in educational opportunities with 30 elementary schools, six middle schools, nine high schools, two universities, and a community college. All area high schools provide vocational training in addition to basic education, and there are work-focused internships as well. The reported public school enrollment in the NRV is 20 041 (1997–98).[59] Radford University, located in the City of Radford, is a four-year, state supported university serving over 8500 undergraduate and graduate students. Virginia Polytechnic Institute and State University, located in the Town of Blacksburg, is a four-year, coeducational, comprehensive land grant university with an enrollment of approximately 25 000 students. The curriculum offers graduate and undergraduate programs in agriculture, business, engineering and science. New River Community College with an enrollment of 4500 is a two-year non-residential state institution of higher learning that offers a comprehensive spectrum of occupational and technical education, continuing adult education, special training and development programs, as well as specialized regional/commercial services.

In the 1970s, the area's leaders advocated the need to diversify its economic base by leapfrogging from an agricultural to knowledge-based economy. It was thought that the area's resources could be better utilized by developing a value-added knowledge industry using local natural resources and highly trained manpower. Specifically, the idea of university technology transfer leveraging Virginia Tech's resources offered numerous opportunities. The next paragraph provides an overview of the NRV region's key infrastructure elements that enable the region to support its main innovation pole described later.

Regional technology infrastructure

This rapidly developing modern innovation pole offers several of the modern infrastructural elements including a solid science park and several emerging incubator programs, along with trained manpower and university R&D results. Not only are new firms based on university technology being nurtured, but outside firms are also being solicited through generous relocation packages. The key players active in the development of a regional technology-based economic development milieu around Virginia Tech comprise the State of Virginia, Virginia's New River Valley Economic Development Alliance, the surrounding Montgomery County, and the Chamber of Commerce representing local private businesses.

Mobilizing its university and regional resources, the NRV area has developed an enabling technology infrastructure that attracts both fledgling technology-based firms and those national and foreign firms planning to relocate to the region. Being a rural area, away from the traditional centers of business activity, the region's seed and risk capital industry is underdeveloped, and efforts are underway to improve this.

Given Virginia Tech's strengths in electronics, computer and information technologies, and biotech, the area offers a broad spectrum of technologies. Therefore, there is no special emphasis on a single or a set of sectors other than being high-tech. Outside Virginia Tech, the nearby local communities have set up at least three incubators with help from the federal Economic Development Agency and state money. In the adjacent Montgomery County there is a technology incubator/multi-tenant facility (30 000 square feet or 2787 square meters) housed in one of the industrial parks. There is a functional airport strip next to the university facilities. In short, the NRV region has endeavored to strike the right balance in setting up of the technology development mechanisms and related infrastructure elements. As part of the technology infrastructure development Virginia Tech Corporate Research Center (CRC) was established in 1985. Currently, there is a virtual incubator called Business Technology Center that serves clients from throughout the state. A new business accelerator is being established at CRC. The following paragraphs provide a detailed case study of the Virginia Tech Corporate Research Center as a main innovation pole of the region.

Virginia Tech Corporate Research Center

Situated in the foothills of the Blue Ridge Mountains of the State of Virginia, the Virginia Tech Corporate Research Center (VT-CRC) is a university-related science park. The physical facilities are spread around 125 acres (over 50 hectares) of land adjacent to the Virginia Polytechnic Institute and State University (also known as Virginia Tech) in the town of Blacksburg, Virginia. Founded in 1985, the Center currently has 19 buildings with 675 000 square feet (62 708 square meters) space for lease, all of which are wired with broadband telecommunication facilities.

Another salient feature of the park is its location next to the Virginia Tech Airport which is half an hour by road from the main Roanoke Regional Airport, offering a uniquely accessible small town rural environment. The CRC continues to expand with multiple buildings in planning and construction phases. There are more than 120 tenant organizations, the majority of whom (around 85 per cent) are private firms, employing 1830 people including 300 Virginia Tech students. Most of its tenant companies are national with only a couple having an international business clientele. There is a good mix of technologies and the park is technologically as heterogeneous as it can be.

The Center's (CRC) mission[60] is to help develop technology-based firms while advancing Virginia Tech's teaching, research, and outreach missions. The area offers research strengths and facilities of the university; reduced costs by using high-quality capable student labor; support for companies from the CRC's programs; and a good quality of work and home life that this region affords. The tenant companies can avail themselves of comprehensive business assistance programs, a modern telecommunications infrastructure, a financial assistance package, training opportunities, and many benefits that result from the CRC's affiliation with Virginia Tech, such as faculty privileges, numerous recreational and sports opportunities, and a restaurant.[61]

Organizationally, the CRC is a wholly, owned for-profit subsidiary of Virginia Tech Foundation. As a private corporation, the CRC is independent from the university. Hence it offers a private sector environment but with access to the resources, facilities and expertise of Virginia Tech, a major public university. Table 2.9 provides summary data on the VT Corporate Research Center facility.

Park history and origins During the late 1970s several Virginia Polytechnic and State University leaders deliberated on the issue of leveraging university resources to participate in the economic development of the region. In the early 1980s, the state government was also exerting pressure on the university to help the state business community in promoting entrepreneurial activity. Virginia Tech is the Commonwealth of Virginia's largest land-grant university. The university was founded in 1872 with a mission to provide education, to

Table 2.9 Virginia Tech Corporate Research Center: summary data

Sponsor(s)	Virginia Polytechnic Institute and State University
Year opened/settings	1985/ Rural
Size: Land/buildings	125 acres (50 hectares)/19 buildings with 675 000 sq. ft. (62 708 sq. m.) space
Start-up support (Incubator)	Virtual incubation function through the 'Business and Technology Center' located in the park
Affiliated organizations	VT Intellectual Properties Corporation, Venture Capital Bank, Small Business Development Center, Pamplin School of Business Entrepreneurship program
Resident organizations including TBFs	Park: 121
Tenant employees	About 1830
Specialty/focus	Technology transfer and commercialization of a broad range of university technology
Mission	To promote university technology transfer and opportunities for faculty and students. Pursue area economic development through biotech, IT and other firms
Organization and governance	The park is a not-for-profit subsidiary of VT Foundation. Governance is through a board with wide regional representation
Financing, capitalization, revenue	VT Foundation, State of Virginia support
Marketing – targeted technologies and type of entrepreneurs	About half of the tenant firms are computer/internet related, others include biotechnology; entrepreneurs are equally divided between faculty, local, and out-of-state
Tenant services	Business and technology development services, seed capital, VC access, library and databases, university facilities, manufacturing not allowed in the park. Nearby industrial park available for relocation
Technology-based firms surveyed	A total of 55 TBFs were identified for the survey and 13 responded (24% response rate)

Source: Corporate Research Center Survey

advance knowledge through research and enhance quality of life for the people of Virginia. As a land-grant institution, outreach in the form of public services is an important part of its mission and upon public demand in 1982 a university committee was constituted to look at this aspect of the mission. In 1984 the consulting firm Peat Marwick was also engaged to help study the issue, which recommended establishing a technology transfer company, and a research park. The result of these deliberations was the recommendation that a science park be established to help create a platform in technology transfer from the university to the private sector. Hence, the idea of Virginia Tech Corporate Research Center (VT-CRC) was conceived. The park was established as a wholly owned subsidiary of the Virginia Tech Foundation in 1985. A 120-acre (48 hectares) piece of farmland was secured next to the university airport. The Foundation provided $4.2 million cash to build the infrastructure facilities. The Center received an initial dose of public money of $600 000 from the Economic Development Assistance (EDA) program of the US Department of Commerce, which was also utilized for infrastructure development. The Foundation borrowed $30 million to build various building units. However, up until the time of our survey, when the facility was estimated to be a $35 million investment, there had not been any more government funds invested after the initial EDA grant.

Undoubtedly, the primary resourceful entity for the CRC in the region is Virginia Tech, which is ranked among the top 50 universities in the USA in terms of total dollars expended on research, and is among the top 10 per cent of those universities having significant amounts of industry-sponsored R&D funds. The university has a history of interdisciplinary research, which has led to the creation of research centers of international renown in many areas such as adhesives, biotechnology, composite materials, computer-aided design, fiber optics, environmental science, energy research, polymers and power electronics. The university, located in a small rural town known as Blacksburg (population 50 000 including 26 000 university students) has a full-time instructional faculty of 1424, 5000 graduate students and 20 500 undergraduate students. It offers about 70 bachelor degree programs, and 120 masters and doctoral programs. Virginia Tech's budget is approximately $500 million annually (2000), with more than $30 million received annually from private gifts. Alumni, totaling 130 000 located in over 100 countries, provide gifts, which are used to fund the Virginia Tech Foundation. The university conducts a $165 million-research program each year that is associated with more than 3500 projects in its hundreds of research laboratories.

To provide business assistance to new start-ups, a business and technology service was contemplated back in the early 1990s. A person with a PhD in chemistry was retained on a part-time basis for this purpose. His job was to meet 4–5 companies a year and to find new business opportunities, and typically 1 or 2 companies would come out of this effort – a boutique operation. In 1994 the people from the Center for Innovation Technologies (CIT), a state entity, met with the CRC president and the dean of Virginia Tech's Pamplin School of Business to enhance the scope of this activity and restructure the operation. As a result, the VT Business/Technology Center (BTC) was created. The center is physically located in the CRC but it operates all over the Commonwealth of Virginia. At the time of the survey, it was reported that there were on average 100 clients and 200 projects with which the BTC works in a year.[62] Up until recently, the CRC did not have a formal incubator; the BTC provided some services and helped new start-ups on the business/technology assistance advice side.

The Virginia Tech Corporate Research Center plans to establish a 45 450 square foot incubator to launch high-tech companies. The $5.8 million project is expected to commence by 2005 and will be partially funded by a $2 million grant from the US Economic Development Administration; the Virginia Tech Foundation will pay for the rest. The goal is for the incubator enterprises to graduate to their own facilities in the CRC or to industrial parks in Montgomery County or elsewhere in the New River Valley.[63]

Characteristics of tenant firms The CRC tenants represent the following technologies: biotech, computer science, diagnostics, electronics, engineering, environmental, internet/telecom, library science, materials and chemistry, transportation, technology transfer, agriculture, and others.

During the survey, the park manager revealed that 'over the past eleven years the park had a total of 133 tenant customers, 87 being still tenants at the time of the study. Of the graduates, about 50 per cent have simply relocated, 17 per cent have failed, and 10 per cent were acquired by others generally at the Park'. 'In 1999, the park's 87 technology-based firms reported a total employment of 1320 out of which 43 per cent were university related, alumni, faculty and students.' 'The number of tenant firms and their employment over the past several years has grown steadily.' 'One third of park entrepreneurs came from the university (mostly faculty).'[64]

The greatest number of tenants have been in computer science. Other technology areas with significant representation are internet/ telecommunication, electronics and design, chemical and materials, environment, and biotech. For the purpose of this study, out of the total 87 CRC tenant organizations, 55 firms that fall in our category of technology-based firms were targeted. However, only 16 firms, mostly computer science-related, participated in the survey (18 per cent return rate). The data shows that with the exception of two firms (established in 1981 and 1989) all of the other responding firms were established in the 1990s with most in the first half of the decade.

Over the past five-year period, about 23 product (both goods and service), nine process, and six management-related innovations were reported (about two innovations per firm on average). Eleven respondents said that market pull was the key driving force of these innovations, while 10 said it was technology push. Another nine expressed the roles played by both (equally by market as well as technology push) as the contributing factors. In terms of sources of innovation, a majority (69 per cent) of the respondents said that internal sources were responsible for the success of these innovations. Ten trade secrets, seven patents and six designs were reported as being used by the responding firms. Most of these firms are national corporations, with a smaller number of single privately held ownerships; only one of the responding firms was a foreign subsidiary. The market coverage of their products/services was reported to be national in most of the cases, with the exception of three cases with both national as well as international coverage. Though some technology-based firms surveyed refrained from sharing their sales data, citing privacy reasons, those firms who shared this information reported generally positive trends in their sales figures over the past five years, showing them to be successful operations. A review of the percentages of the workforce engaged in R&D put together show an average of almost three times as many personnel engaged in R&D than in other areas of business (sales, marketing, management and so on).

This analysis shows a typical profile of a technology-based firm being successfully nurtured in a science park/innovation center setting, in this case the CRC facility. Though computer science-related companies draw most faculty entrepreneurs and have the least attrition rate, biotechnology and agricultural technologies make up a significant part of the park. Among the examples of success stories out of the Center [65] are tenant companies like VTLS Inc., a global library information systems provider; Tech Lab Inc., a microbiology and medical diagnostics research firm; Revivicor, Inc. a medical research company for the replacement of cells, tissues and organs; and Recognition Research, a new software technology and design systems firm for document imaging.

Performance and regional impacts A review of the past more than one and a half decades of the CRC operation shows a continuing growth along with a number of tangible outcomes.

The Park was created as a response to a combination of the business community's request and the Virginia Governor's wishes. The purpose was to connect the business community to the university resources and to connect all the pieces within the university and the region that relate to economic development. The intention was also to facilitate partnerships to help technology transfer. The CRC's ability to market itself as something other than a real estate venture has been remarkable. The union between the CRC and the university serves the purpose of forming lasting partnerships, pursuing unique research, and developing valuable new products. Since the formation of the CRC, sponsored research at the university has been growing at an annual rate of over 8 per cent. An increasing proportion of this can be attributed to the existence of the CRC. Thirty-three per cent of the CRC tenants sponsor research at Virginia Tech. The level of sponsored research from companies at the CRC to the university was more than $1.2 million in 1998. The CRC continues to be a visible example of Virginia Tech's commitment to research excellence and to economic development in Southwest Virginia. The CRC has fostered the growth of over 22 new companies based on technology developed at Virginia Tech. Additionally, faculty, staff, and students maintain an active association with companies in the park. The CRC prides itself on being very closely linked with university personnel.

As mentioned above, there are over 120 tenant organizations employing 1830 people with a combined payroll of more than $8 million. These are mostly high-quality, high-wage jobs that have a positive impact on the overall economic health of the community. Four companies have left the park to carry out manufacturing and have created over 200 additional jobs in the regional and local economy thus far. The CRC is also experiencing growth in the business support infrastructure providers, that is, lawyers, accountants, and other professional consultants.

In 1999, the Virginia Tech Corporate Research Center received the distinction of being given the award for 'Best Practice in Technology Transfer and Research Centers' by the National Council for Urban Economic Development. Research centers in eight states from Georgia to Pennsylvania were considered for the award.[66]

Conclusion The New River Valley (NRV) Region's burgeoning population owing to its favorable climate and central location, together with the presence of a major research university with R&D-intensive environment, prompted area leaders to plan for economic diversification. As part of the innovation infrastructure development VT-CRC was established in 1985. Today, this rapidly developing modern innovation pole offers an expansive, highly developed infrastructure, attracting well over 100 tenant organizations, mostly technology-based firms and support organizations and 1830 employees, including a large number of Virginia Tech students as part-time employees and interns. The tenant companies are managed conservatively with emphasis on high potential technologies at the time when the CRC and and its surrounding infrastructure are expanding. Technology clusters are developing in biotechnology and transportation. Larger companies are also being attracted to the park. The research base of the university continues to grow at about 20 per cent per year. As a result, the CRC is becoming recognized as the center of technology development in the state and beyond.

Leveraging its university R&D strengths and regional resources, this area is well on its way to developing a state-of-the-art innovation infrastructure that supports indigenous TBFs at the same time capturing and retaining outside firms that are attracted by its location. However, the area's venture and seed capital industry is underdeveloped and efforts are underway to improve this. There are plans to add a state-of-the art incubator at the CRC facility to help commercialize university technology by launching high-tech companies. After the completion of this project, CRC can claim to have most of the elements needed in a comprehensive incubation space providing seamless innovation opportunities for technology-based enterprise development in the New River Valley region.

2.2.4 Wisconsin's Madison Metropolitan Area

The Madison Metropolitan Area and its environs are spread over an area of 1230 square miles (3186 square kilometers), with 35 lakes. This area falls in Dane County and consists of 48 municipalities outside the City of Madison including five small cities, eleven villages, and 32 towns. The area is easily accessible through all major highways and several railroad systems with Chicago only 142 miles to the Southeast. It is served by more than 40 common carrier truck lines and major bus services. The Dane County Regional Airport is served by major airlines and offers full passenger and airfreight services with non-stop flights to key regional and national destinations.

The 2001 population statistics show that out of 431 815 county inhabitants, 210 377 (48.7 per cent) live within the City of Madison with the remainder living in the adjoining towns, villages and small cities. Dane County's total non-farm wage and salary employment of 295 580 (2001) is divided as: 34.8 per cent service, 25.6 per cent government, 21.2 per cent wholesale and retail trade, 14.8 per cent manufacturing including construction and mining, and 3.5 per cent transportation and utilities. Only 1290 individuals (0.4 per cent) were associated with farming in 2001 – a considerable decline over the last three decades. Dade County's per capita income (in 1997) was $27 361, which is higher than this farm state's $24 048 average (the US average in 1997 was $25 288).[67]

The following paragraph provides an overview of the Wisconsin Madison Metro Area's key infrastructure elements that enable the region to support its main innovation pole described later.

Regional technology infrastructure

Sixteen public school districts and 22 public schools with a total enrollment of

24 943 students serve Dane County. There are seven institutions of higher education with a total enrollment of 73 790 students. The University of Wisconsin-Madison (UW, with 40 109 undergraduate and graduate students) and Wisconsin area technical colleges (with 30 078 students) – make up the bulk of higher education enrollment.

Although government and service employment is more than half of the total, the technology sector represented by biotechnology, medical/biomedical research, microelectronics, software or other computer-related firms dominate Dane County's research community; more than 375 high-tech businesses employ more than 20 000 people (7.4 per cent of the total employment).[68]

The UW's history of involvement with patents and a technology transfer program through its Wisconsin Alumni Research Foundation (WARF), and the pressure from shrinking public financial support for the university, prompted the establishment of an innovation infrastructure led by the university. It was hoped that this will help in aiding commercialization of university research for future financial gains.

Cognizant of the UW-Madison's R&D strengths, a highly educated workforce coupled with the area's high quality of life, the university officials started deliberating on the development of technology infrastructure during the early 1980s. The University Research Park facility was established in 1984.

The area's main source of technology, UW-Madison has strengths in several technology areas including biotech and information systems. However, biotech/medical and services make the bulk of technologies represented in the area's science park. The area is consistently ranked as one of the best regions in the nation in terms of quality of life amenities.

The region is a highly attractive destination for high-tech business relocation and new start-ups. With the help of the area's utility company, Madison Gas and Electric (MG&E), a full-blown incubator is also functional. There is already a well developed technology and business development support infrastructure in place. Seed and VC help is being made available. With the help of UW-Madison's leadership, the area has successfully developed a modern and growing technology and business development infrastructure with its science park as the center piece that has earned recognition. The future expansion of the park facility and continued reliance of active university involvement has its limitations, which may pose challenges in the future. The following paragraphs provide a detailed case study of the University Research Park, Madison as the main innovation pole of the region.

University Research Park, Madison

The University of Wisconsin-Madison's (UW-M) University Research Park

(URP) is a university-related research park located in suburban Madison, three miles away from the UW-Madison campus. The URP is a unique and diverse science park encompassing 351 acres (142 hectares) of original parkland of which 255 acres (over 100 hectares) is developed. There are 34 buildings with 1.5 million square feet (139 350 square meters) of space. It is home to over 100 of Wisconsin's mostly high technology companies including more than 30 start-up firms. There are a total of about 4000 employees working within the University Research Park.

The UW Board of Regents established University Research Park Inc. (URP) in 1984 as a separate non-profit organization. URP is responsible for the land development and leases to private firms and other park tenant organizations, and provides start-up support to the tenant entrepreneurs through its incubator – the Madison Gas & Electric (MG&E) Innovation Center.

Though the original vision of the park was to focus on bioscience firms, there are a host of other technologies also represented among tenant firms. The Park's primary mission was described to be a mechanism for university technology transfer, but it provides an atmosphere custom-designed to nurture a productive combination of economic and technological development.[69] Table 2.10 provides the summary data on the University Research Park case study.

As shown, the park supports an innovation center that houses technology-based start-up companies through its comprehensive business development network. The center provides an environment tailored to the needs of companies seeking to improve their competitive position via timely commercialization of new technologies. The various complementary organizations located in and around the park serve the discrete needs of companies at every stage in the product development process. These components include: Science Center, MG&E Innovation Center, UW Graduate School, Wisconsin Alumni Research Fund (WARF), University Industry Relations (UIR), and many funded interdisciplinary research centers that have no direct relationships with URP. There are several research consortia build around 20–25 fields of research (with membership costing several thousand dollars each) in which private firms participate and underwrite general research.

Tenants firms 'graduate' from the MG&E Innovation Center when they reach the ability to function on their own. Most Innovation Center companies have relocated either to other park buildings or outside the research park. Besides the core technology-oriented organizations and firms there are other support organizations providing banking, accounting, venture capital, and engineering and business law services to the client firms. The other park amenities include childcare and pre-school facilities.

Park history and origins Intrigued by the growing science park movement in

Table 2.10 University Research Park, Madison: summary data

Sponsor(s)	University of Wisconsin, Madison
Year opened/settings	1984/Suburban
Size: Land/buildings	Original 351 acres (142 hectares) land with 255 acres (100 hectares) developed/34 buildings with 1.5 million sq. ft. (139 350 sq. m.) space
Start-up support (Incubator)	Located in the park 'MG&E Innovation Center' with lease area of 36 000 sq. ft. (33 444 sq. m.)
Affiliated organizations	Science Center, University Industry Relations, Wisconsin Alumni Research Foundation, University Graduate School, Office of Technology Commercialization, Entrepreneurship Center
Resident organizations including TBFs	Park: 77, Incubator: 35
Tenant employees	About 2300
Specialty/focus	Technology transfer and commercialization of biotechnology and other areas of technology of university strength
Mission	To promote university technology transfer and opportunities for faculty and students. Area economic development through biotechnology, IT and other firms
Organization and governance	The park is organized as two not-for-profit corporations. Both operate under a board of trustees with university majority. The park director reports to the board
Financing, capitalization, revenue	WARF, State of Wisconsin; The Park has a $76.5 million capitalization and annual budget of around $6.17 million
Marketing – targeted technologies and type of entrepreneurs	Two thirds of the tenant firms are high-tech, and half of the firms have university ties including faculty and alumni entrepreneurs
Tenant services	Business and technology development services, seed capital, VC access, and most of the university facilities, R&D, assembly and light manufacturing allowed
Technology-based firms surveyed	35 TBFs were identified for the survey out of which 7 responded (20% response rate)

Source: University Research Park Survey

the nation and cognizant of the economic development mission of his prominent land-grant institution, in 1979 UW-Madison Chancellor Irving Shain appointed an experienced state employee Wayne McGown as Special Assistant to the Chancellor to study and implement the idea of a university-related science park that would serve as a focal point for promoting technological entrepreneurship in the region. During this period, it was felt that Wisconsin is a case in the rust belt where any economic revival would require leveraging the knowledge sector and realizing global connectivity. The university was educating a lot of PhDs who tended to move out of state for jobs in more technology-oriented areas in the nation.

It was not until 1981 that the idea of a science park got on the drawing board with Wayne McGown working closely with several interested stakeholders from the government and the local private sector. A 1982 study by the Urban Land Institute of Washington DC concluded that the strong research emphasis of the UW-Madison campus and a nearby 325-acre (132 hectares) parcel of the university's agricultural experimentation land that was no longer suitable for agricultural research purposes owing to urbanization was ideal for the creation of a university related research park. The concerned university leaders along with the City of Madison officials and area business leaders made study trips to some of the prominent examples of science parks in the nation and made a positive recommendation to establish a park in the area. After developing a consensus as a result of satisfying all facets of the community, the UW-M Board of Regents approved the establishment of a University Research Park (URP) in 1983. A joint action by the Board of Regents and the State Building Commission in 1984 gave permission to University Research Park Inc. to develop the land. The same year the university sold the land to URP. The Warzyn Company, a traditional engineering company of long standing in the community became the first park tenant in 1985. Persoft, a computer software company, followed as the second park tenant.

Soon it was realized that the initial model of large companies building for themselves did not cover the whole market. According to the park director, 'as we really looked at the experience and analyzed what was going on, we saw there weren't a lot of large companies at that point, but there were a lot of small early-stage companies that were interested ... and those companies needed either office or laboratory space.'[70] This realization led to the next major milestone of the park – the Science Center Buildings – a multi-tenant complex of smaller amounts of space for smaller firms. It was decided to reserve one of the nine Science Center Buildings for start-ups only. Therefore, the Science Center Complex of nine buildings, one housing the new innovation center, was completed during the 1989–92 period. The challenge of financing such speculative spaces was overcome by soliciting support of the local utility company, Madison Gas and Electric Company (MG&E) which had an active interest in the area's economic development. Over the next several years MG&E provided $400 000 to establish the building specified for startups inside the Science Center Complex. A $250 000 state grant helped develop the laboratory space in the Science Center. Later, to meet the growing demand for high-tech business start-ups space in the park region, the MG&E donated another $1 million toward the new MG&E Innovation Center building in the park.[71]

The companies that grow out of the Science Center or MG&E Innovation Center start-ups, which require more space or new tenants from outside the park may choose to construct their own facilities on parcels leased from University Research Park, Inc. As the University Research Park nears capacity, plans are being developed for the future. Recently the Board of Regents of the University of Wisconsin System negotiated the purchase of 114 acres (46 hectares) of land in the nearby town of Middleton for research park development. According to the plan, the land will be used for a second University Research Park once the current park reaches full capacity.[72]

Characteristics of tenant firms At the time of the survey,[73] out of 102 park tenants about one third (31 firms) were service-related (business, financial, medical), another one third (32 firms) were evenly divided into biotechnology and information/software-related firms, and the rest were mostly university-related services, including health facilities such as the Research Park Clinic, University Community Clinic, and a portion of their administrative services. The UW-Madison Psychiatric Institute and Clinics and the University Health Care are also located in the park. It can be said that the University Research Park is the picture of diversity housing, with not only science and technology firms but also many other types of facilities that enhance the support for science and technology research. The rough percentages of various technologies represented in the park show the biotech/medical area as having the largest share (37 per cent) followed by service firms (31 per cent). For the purpose of this study a population of 32 firms falling into our category of technology-based firms was targeted. Only 11 firms (31 per cent return rate) including start-ups participated in the survey.

The survey shows that with the exception of one firm (established in 1989) all of the other responding firms were established in the 1990s with most in the first half of the decade. Over the past five-year period, about 19 product (both goods and service), five process, and four management-related innovations were reported (more than three per firm on average). Thirteen of these innovations attributed market pull as the key driving force, while only four to technology push, and another 11 innovations had roles played equally by both (market pull as well as technology push).[74] In terms of sources of innovation a majority (77 per cent) of the respondents said that internal sources were responsible for the success of these innovations. A majority (74 per cent) of the respondents reported making use of patents, and to a lesser extent they used trademarks and design respectively. Most of these firms are locally or nationally owned, with a smaller number of single privately owned businesses; only two of the responding firms had a foreign share in their ownership. The market coverage of their products/services was reported to be national in most of the cases, with the exception of only two cases that had both national and international coverage. Though most of the technology-based firms studied refrained from sharing their sales growth data citing privacy reasons, the firms who shared this information reported positive trends in their sales figures over the past five years (1995–2000), proving them to be successful operations. A review of the percentages of the workforce engaged in R&D put together shows an average of two and a half times as many personnel engaged in R&D than in other areas of business (sales, marketing, management and so on).

This analysis shows a typical profile of a technology-based firm being successfully incubated in a science park/innovation center. However, the key limitation in the generalizability of these results is the fact that while two thirds of the tenants are located in the science park and one third in the MG&E Innovation Center, only a couple of the eleven firms surveyed are from the MG&E Innovation Center. Some of the prominent corporate tenants in the park include the following: Ultratec, one of the longer term and largest high-tech employers in URP; it represents the type of companies that serve as anchor tenants. Once in a facility which was large enough, Ultratec brought its manufacturing in-house and grew from less than 20 researchers to a staff of over 150. PanVera Corporation, a graduate of its own Innovation Center is a successful manufacturer and marketer of recombinant proteins and fluorescent assays for use in health care research. Tetrionics, an organic pharmaceutical synthetic lab also spent seven years in the Innovation Center before moving to an independent building in the park. Another high-tech firm Novagen, which was also in the MG&E Innovation Center and now is housed in the Science Center of the park, manufactures and sells molecular biology reagents and kits.

Performance and regional impacts A review of the overall performance of University Research Park shows a clear success. The partnership of a prominent state university and the local private sector was instrumental in developing university–industry relationships by creating a new technology sector. The region benefited by transforming 255 acres (103 hectares) of marginally used land into a tax-paying and job-generating cluster of new firms.

Over the past 20 years, the park has helped develop a new economic base for the City of Madison and its environs, adding its technology businesses to health care, information technology, business services, and other sectors as the local businesses and employers of the future. To keep the momentum, the park is planning to acquire nearby new private land toward the west side of Madison.

On the technology side, the park now houses more than 100 organizations including high-tech firms and new start-ups, employs over 2300 mostly high salaried professionals and has dramatically impacted the economic base of Madison as well as enhanced the university's image. UW-Madison students and faculty have enjoyed collaborations with Research Park companies through internship programs, company creations, and research partnerships. Today, several faculty members and students and/or alumni have founded companies in the park, and most importantly, research park companies have provided active learning laboratories to augment the curriculum in programs at the university's various schools. These combined resources have led to the creation of many full-time as well as part-time jobs for the university community.

Beyond the bounds of Madison and its vicinity, the Research Park has drawn the attention of delegations from around the world. Visitors interested in research and technology transfer, economic development, university–industry relations, and business incubation travel from all around the nation and the world to learn from park staff what works and what may not work in the creation and development of research parks.

According to an estimate, UW-Madison received a positive return on its investment in URP – during the 1982–98 period the return (in the form of rents, equity and so on) was reported to be $10 647 000 against an investment of $4 280 000 for the same period.[75] Another goal of the Research Park development has been to leverage taxes for the City of Madison – in 1999, the Research Park paid $1 535 000 in real estate taxes and in 2001, the tax revenue exceeded $2 million per year; the total revenue collected during the 1990s was more than $12 million.

Operationally, the park is largely self-supporting with 81 per cent of the expenses covered from rental income, 13 per cent from land rent and the remaining 6 per cent comes from interest income – the university does not provide any financial support for operations. According to a recent estimate, over the past fifteen years, the incubator tenants and graduates have generated more than 1500 full-time jobs and annual gross revenues in excess of $120 million. In the late 1990s, the incubator tenants and graduate firms employed 665 full-time equivalent employees and paid around $622 000 in taxes.[76]

To recognize this success, in 1996 URP received the first 'Outstanding Research Park Achievement Award' from the Association of University Research Parks – chosen from a field of over 144 research parks throughout North America.

Conclusion Twenty-five years ago, under UW-Madison leadership it was envisioned to help develop a new economic base for the Greater Madison Area through an entrepreneur-driven knowledge economy by leveraging the university's strengths in several technologies. This resulted in the establishment of URP in 1984. There are now several success stories associated with this initiative, a cluster of over 100 firms and other knowledge-generating organizations with 4000 knowledge workers in the park, a sophisticated support infrastructure, a private sector-funded innovation center, and increasing seed funding and VC help among others. This suburban science park, which has a maturing incubator, is now a focal point in mobilizing university-developed technology for commercialization and a highly attractive destination for high-tech business relocation and new start-ups. There is already a well developed technology and business development support infrastructure in place and the area is increasingly being recognized as an emerging knowledge region.

URP is now considered a viable park with a potential to generate income from the university developed technology. The focus has recently shifted toward figuring out growth patterns for the future, being more responsive to tenant needs, and creating more institutionalized vehicles for faculty involvement in commercialization that will potentially begin the process of creating an income stream for the university. It appears that the sustainability and growth of the park facility is contingent upon a continued reliance of active university involvement in the future, which has its limitations. Therefore in order to ensure future viability of the region as a TIP, especially in developing a local venture capital industry or expanded park facilities, involvement of the private sector in such areas seems to be the logical course.

2.2.5 Summary and Conclusions

Starting with each knowledge region as a unit of analysis, the above four cases provide insight into the ways in which their main science parks and affiliated incubators are contributing in developing them as successful technology innovation poles. The regional milieus found in all of the four cases have several characteristics that signify them as vibrant TIPs (or developing as such) which provide ample opportunities and an enabling environment for promoting innovative firms in their respective regions. In spite of their own unique characteristics and historic strengths in various fields of technology, all of these regions had somewhat similar preconditions in terms of socio-economic profiles prior to their development as TIPs. In all these cases, there appears to be a common thread in terms of developed higher education institutions, historically active private industry, and supportive government sector providing incentives for restructuring. These efforts were often undertaken as partnerships among academia, industry, and government, though the extent of involvement of each of these factors differed from case to case. The presence of committed champions among the key factors along with the land-grant mission of the state universities played a significant role in these transformations.

In Virginia and Wisconsin the state governments' encouragement of their main state universities to get involved in developing an incubation mechanism to support university technology transfer and commercialization activities resulted in the establishment of VT-CRC by Virginia Tech and URP by the University of Wisconsin-Madison. On the other hand, in the New York Capital Region and the Chicago Metro Area, lead roles were played by their prominent private universities, Northwestern University in the case of NU/ERP and RPI University in the case of RTP, though the local city government of Evanston was an equal partner with Northwestern during the NE/ERP's initial formative years.[77] In the case of URP, Madison Gas and Electric Company's key role in establishing the innovation center at its park is a salient example of the private sector's involvement with innovation infrastructure development. General Electric's role in generating a spin-off activity in the New York Capital Region is another example of the participation of a large private company in NTBF development.

In both the New York Capital Region and the Chicago Metro Area, there is a visible presence of entrepreneurial culture, qualified manpower, quality of life amenities, a traditional industrial base and developed regional infrastructure (telecommunication, transportation and so on). The rural New River Valley Region lacks a traditional industrial base, yet a rapidly developing regional infrastructure and entrepreneurial culture along with the relatively lower cost of doing business mediates some of the disadvantages. Similarly in the Madison Metro Area, with its moderate level in industrial base and a developing entrepreneurial culture, the university's prominent research activities coupled with superior quality of life amenities also tend to offset the disadvantages. As vibrant technology innovation poles all four park facilities are regionally as well as globally well connected with a visibility that adds to their sustainability and growth.

Comparative reviews of the summary data and profiles of the four park facilities acting as innovations poles reveal the following.

In terms of *sponsorship* all four facilities have partnership arrangements involving university, government and private sector – all these cases have research universities, which are also producing new knowledge through their numerous R&D centers and research laboratories. All of these facilities have active technology business *incubation* programs that provide a supportive environment for nurturing new technological ventures, and a capturing tool for their parks – both RTP and NU/ERP have NBIA's national award-winning model incubator programs, URP has a private sector-funded relatively new incubator program which is growing, and VT-CRC is in the process of developing one. These incubators are also surrounded by several other *affiliate organizations* providing technical and business advice, laboratories and workshops, seed and venture capital connections, databases and internet sources, and entrepreneurial education and training programs.

In terms of goals, like most science and technology parks they seek participation in the regional economic development activities by supporting the development of technology-based firms; providing a laboratory for learning entrepreneurial skills; and promoting commercialization of university technology. Our study of these parks with respect to their past accomplishments shows that all four have made considerable progress in meeting these goals.

Organizationally, with the exception of NU/ERP which is for-profit, all the others are not-for-profit organizations. They all are governed by multiple stakeholders drawn from the university, regional and/or state government entities, and private industry. Incubators generally are incorporated as not-for-profit entities. In three out of the four cases, URP, VT-CRC and NU/ERP, the incubators are not-for-profit and the fourth, RPI Park along with its incubator program is a division/department of the private university. Most of the affiliated organizations are inside or in close proximity to the university and are surrounded by complementary R&D institutes and industrial research laboratories, which are pursuing numerous technology and business development initiatives.

In terms of *targeted technologies*, the new and emerging fields including software, biotechnology, informatics, and electronics firms represent the largest number of tenants; however, the relative marketing emphasis varies according to university strengths and/or regional developmental policies. The participation of university faculty and students as entrepreneurs varies from facility to facility and is generally encouraged by all four facilities.

In terms of *financing* to establish or expand, all of the facilities have received capital funds from a combination of public and private sources. For tenant firms the support for the provision of easily accessible seed and venture capital from multiple sources has been the hallmark of these parks and their associated incubator programs. Moreover, a host of state and federal grant programs are available for which ample guidance and support is provided by the parks to their technology-oriented tenants. As a result, these four parks have a significant percentage of tenants supported through external funds. The rural New River Valley Region and the relatively less urban Madison Metropolitan Area have fewer developed venture and seed capital industries.

In the provision of *services*, all of the four facilities, like most successful parks and their associated incubators, have been responsive to client needs and perceived usefulness of the gamut of services often provided through these mechanisms – shared space, typical office services, conference room and other maintenance services. Previous research shows that technology-based client firms have consistently given higher ranks to the university-related services/benefits, such as university image; use of student employees and faculty consultants; and access to libraries and laboratories (Mian, 1996). Therefore, all four cases provide these services/benefits, depending upon their overall reputation and commitment to technology incubation. Research results on the value-added contribution and, hence, desirability of typical park/incubator services are mixed. However, most of the typical incubator services, including facilitating networking, business and legal consulting, are available in one form or another to most incubator clients.

The outcomes of a successful technology innovation pole are manifested in the form of creation and growth of TBFs and innovations and other related regional impacts as part of the organizational missions in the four park cases included in the study. The factors that mitigate against a developed region's ability to have a significant economic impact may include the region's relative stagnation in population growth resulting from a lack of economic opportunity and/or harsher climate. But these factors were found to be pretty much neutralized by the generally more disciplined labor force with superior work habits as was reported by several of the entrepreneurs in the New York Capital Area and Chicago Metro Area during the interviews. On the other hand factors mitigating against the relatively less-developed regions in their ability to allow technology-based economic development activity to a fuller extent are the lack of (or currently developing nature of) some key infrastructure elements, especially the unavailability of venture and seed capital as mentioned earlier.

Of the four parks and their incubation facilities studied here some have more than two decades of experience and others have pioneered the concept. They have served as models for promoting entrepreneurial economies in their regions – they tend to be flexible in learning from their own experiences and adapting to other 'best practices'.

NOTES

1. 'Innovate America', National Innovation Initiative Report, Council on Competitiveness, December 2004 (http://www.compete.org/pdf/NII_Final_Report.pdf).
2. *The Global Context for US Technology Policy*, US Department of Commerce (1997).
3. *The Europa World Yearbook 2003*, 44th edn, Vol II, pp. 4401–19.
4. US Census Bureau, 'State and country fact sheet', 2004.
5. Turner (2002), *The Statesman's Yearbook.*
6. The Europa World Yearbook 2003, 44th edn., Vol II, pp. 4401–19.

7. Ibid.
8. A transition from the old SIC-based codes to North American Industry Classification System (NAICS) codes was completed by the US Department of Commerce in 1999. The resulting list of high-technology NAICS codes includes a total of 39 codes that range from 4 to 6 digits. Twenty-nine of these codes apply to manufacturing industries and ten represent service industries, which incorporates more numerous and broader codes (than the old SIC-based codes) pertaining to rapidly growing industries such as communications, audio and video equipment, and computers.
9. DeVol, Ross (1999).
10. US Department of Commerce (1997).
11. Main Science and Technology Indicators, OECD (2004c).
12. Ibid.
13. National Science Board (2004), *Science and Engineering Indicators*.
14. US Department of Commerce (2004), *The Dynamics of Technology-based Economic Development*.
15. National Science Board (2004), *Science and Engineering Indicators*; OECD (2004b).
16. National Science Foundation (2002), *Science and Technology Indicators*, ibid.
17. Definition of the high-tech industry sector is somewhat controversial. In this study, we are interested in the individual contributions of high-tech industries to the relative innovation potential and hence economic performance of the regions. For these reasons we are focusing on the value of R&D intensity and/or output for industries that may be considered high-tech. Manufacturing industries such as drugs, computers and equipment, communications equipment, electronic components are included, as are service industries such as communications services, computer and data processing services, and research and testing services.
18. National Science Foundation, 16 October, 1998.
19. US Department of Commerce (1997).
20. National Science Foundation (2004).
21. National Venture Capital Association (2005); Science and Engineering Indicators (2004).
22. Thomas Venture Economics, Special Tabulations, Table 6–6, 'New capital committed to U.S. venture capital funds: 1980–2002', National Science Foundation (2004).
23. Wolf (2001).
24. The Morrill Act created a state–federal partnership that diffused the results of a private–public cooperative technology relationship. The Act established the first US technology cooperative program. It created the land grant colleges, whose leading objective was to teach subjects related to agriculture and 'the mechanic arts' or technology. An extension service was created later with the land grants institutions as its nexus.
25. Berglund and Coburn (1995).
26. CRS Issue Briefs for US Congress: Industrial Competitiveness and Technological Advancement: Debate Over Government Policy (IB91132), 5 December, 2000.
27. The federal agencies include: Department of Agriculture, Department of Commerce, Department of Defense, Department of Education, Department of Energy, Department of Health and Human Services, Department of Transportation, Environmental Protection Agency, National Aeronautics and Space Administration, and National Science Foundation.
28. The average number of SBIR awards made during 1998–2000 was 4756. For details see http://www.sba.gov/sbir.
29. The National Science and Technology Council (NSTC) was established by the President of the United States on 23 November, 1993. This Cabinet-level Council is the principal means for the President to review efforts in science, space and technology to coordinate the diverse parts of the federal research and development enterprise. The President chairs the NSTC. Membership consists of the Vice President, Assistant to the President for science and technology, Cabinet Secretaries and Agency Heads with significant science and technology responsibilities, and other White House officials.
30. SSTI Weekly Digest (1997).
31. National Science Foundation (2002).

32. US Department of Commerce (2004), *The Dynamics of Technology-Based Economic Development.*
33. Berglund and Coburn (1995).
34. Betz, Frederick (1994).
35. Berglund and Coburn (1994).
36. Ibid.
37. Mian and Plosila (1996).
38. Ibid.
39. Berglund and Coburn (1994).
40. Berglund and Coburn (1994) and Burglund (1998) SSTI Newsletter.
41. For a comparative case study approach see Yin (1984), Mian (1997).
42. http://www.fiscalpolicy.org/SOWNY2003/CapitalRegion.pdf
43. National Governors' Association (2002).
44. Interviews with several RPI science park entrepreneurs, 2001.
45. http.www.rpi.edu/dept/rtp
46. The TBF mail surveys were administered in early 2000, which followed facility visits and interviews with several TBF entrepreneurs that continued until 2001.
47. For details, see Mian, S. (1997).
48. Interviews with RTP's management personnel, 2001.
49. One of the most notable developments is that SEMATECH (a consortium of 12 international semiconductor companies) is locating an R&D facility with 250 scientists and engineers in Albany. Additionally a high-tech Japanese company, Tokyo Electron Ltd, is opening its R&D center in conjunction with SEMATECH.
50. US Department of Commerce (1997–98).
51. Interviews with the NU-Evanston Park officials, 2001.
52. The TBF mail surveys were administered in early 2000, which followed facility visits and interviews with several TBF entrepreneurs that continued until 2001.
53. AURP (2003), University research park profiles, ARI, MD.
54. Ibid.
55. Interviews with Northwestern University officials, 2000.
56. Ibid.
57. Ibid.
58. http://www.nrvalliance.org/
59. Ibid.
60. http://www.vtcrc.com/generalinformation
61. Ibid.
62. Interview with BTC manager, April 1999.
63. Roanoke Times & World News, 27 August, 2002.
64. Interviews with Park management, April 1999.
65. Interviews with the CRC management and Park tenant surveys, 1999, 2001.
66. http://www.unirel.vt.edu/vtfoundation/report99/toc.html.
67. Greater Madison Chamber of Commerce and Dane County Regional Planning Commission, data compiled from the US Bureau of Economic Analysis and Sales & Marketing Management.
68. Ibid
69. Interview with the Park Director, April 2000.
70. Ibid
71. When interviewed, the MG&E representative cited the long-term benefits that were envisaged by the company in providing financial support to the area's business in their crucial years of inception.
72. A new parcel of land which is 3 to 4 miles away from the existing URP facility has already been acquired and is currently being developed (http://www.universityresearchpark.org)
73. The Madison Metro Area visits including the URP surveys were conducted during 1998–2001.
74. Though a total of 11 firms were surveyed, multiple innovations were reported by some respondents.

75. Interviews with the U-W Madison officials, 2001.
76. Interviews with the Park Director, 2001.
77. The City of Evanston plans to withdraw its 50 per cent financial stake in the park after the end of their current contractual obligation. The university is in the process of putting together a new partnership arrangement.

3. Canada's innovation poles and their role as technology incubation spaces

INTRODUCTION

Canada is known for its large size, northern climate and natural resources. Until recently, its economy was dominated by natural resource activities in mining, oil and gas, agriculture, forestry and related support industries. Services now account for over two-thirds of its economic output and manufacturing for a quarter, advanced technology products and services growing at a fast pace. It is a politically and economically decentralized country, with significant power at the provincial level. Its population, small relative to the size of the country, is spread out along the border with the United States in a number of medium size metropolitan centers. Those characteristics have affected the development of the country's technology innovations poles (TIPs) in ways that may be unique.

This chapter starts with an overview of the socio-economic context of the country and an outline of its innovation system at the national, provincial and regional levels. This is followed by a short description of the country's main TIPs and a more detailed look at four of those poles with special focus on the dynamics of their development and some of the incubation mechanisms that have supported their development.

3.1 CANADA'S NATIONAL INNOVATION SYSTEM

Science and technology activities in Canada have, for many years, reflected its natural resource orientation, its size and its small population. Until the turn of the 20th century, agricultural research and geological mapping were the main research activities. The federal government created the National Research Council in 1916, first to encourage university and industrial research, then to perform research in its own laboratories. Industrial research started slowly in the country. The lack of local research activities was often blamed on the level of foreign ownership of the Canadian industry: important research activities were done in the firm's mother country. Only in the mid-1980s did industry become Canada's main research funding sector, ahead of the federal and provincial governments. Although Canada's research base is now diversified, its structure and leading sectors (telecommunications in particular) still reflect its natural characteristics and history.

3.1.1 The Socio-Economic and Science and Technology Context

Socio-economic overview

With a territory of almost ten million square kilometers, Canada is the largest country in the Western hemisphere and the second largest in the world. Most of its inhabitants live within 200 kilometers of its 5000 kilometers border with the United States. Beyond that, northward isolated communities are engaged in mining, oil or forestry activities, and further north are a limited number of small Arctic communities.

In 2003, Canada had about 31.1 million inhabitants. The average annual population growth rate has decreased, from 1.3 per cent between 1986 and 1996, to 0.9 per cent between 1996 and 2000. Changes between 1986 and 1996 are due to births and deaths (3.87 and 1.97 million respectively), immigration (2.34 million) and emigration (0.44 million), net immigration being equal to the natural population change. Government policy favors immigration, especially for highly skilled and educated professionals and their families, as a tool to sustain population and economic growth. Population growth is quite uneven, with the Provinces of Alberta, Ontario and British Columbia growing the fastest (average annual growth rates of 1.9 per cent, 1.3 per cent and 1.2 per cent respectively since 1996), and Nova Scotia, Saskatchewan and Newfoundland growing the least (0.3 per cent, 0.1 per cent, −1.0 per cent average annual change since 1996).[1]

Canada's population is one of the most educated in the world. The high school completion rate was 81 per cent for the cohort of 19/20-year-olds in 1995/98, and over 50 per cent of its 25/29-year-olds have a college/trade or university education.[2] The geographic distribution of educated people over the country, however, is uneven: only 9.0 per cent of the population over 15 years old have university degrees, certificates or diplomas in Newfoundland, compared with 15.1 per cent in Alberta and 16.8 per cent in Ontario.[3] Differences are significantly larger between high-tech metropolitan and rural areas.

Politically, Canada is a federation of ten provinces and three northern territories. The federal government is responsible for foreign affairs, national defense, immigration, criminal law, banking, regulation of trade and commerce. Provincial governments are responsible for education, property and civil rights, the administration of justice, and matters of a local and private nature. In practice, many domains of jurisdiction overlap, leading to some duplication (industry, commerce, transportation) and occasionally to conflict.

Whereas service-producing industries accounted for about 55 per cent of Canada's economic activity in the late 1960s, they now account for over 67 per cent of the country's Gross Domestic Product. Business sector industries account for 83.5 per cent of GDP. Between 1999 and 2003, real GDP grew at an average annual rate of 3 per cent, one of the highest growth rates among OECD countries.[4] In 2004, growth remained significantly above the OECD average, and it was expected to continue to do so in 2005.

Canada's natural resource orientation is still reflected by the importance of its exports of raw materials and semi-finished products: energy, forestry and agricultural products accounted respectively for 12.5 per cent, 9.9 per cent and 6.5 per cent of all exports in 2000. The country has a large trade surplus in those three sectors: C$39 billion in forestry, C$35 billion in energy, C$9 billion in agriculture.[5] Thanks to the solid infrastructure created through the now terminated Auto Pact with the USA, Canada is a major player in automotive products.[6] Exports of these products account for 23.2 per cent of all exports and 21.3 per cent of all imports, resulting in a large trade surplus of C$21 billion in 2000. Although exports of machinery and equipment (C$107 billion in 2000, 25.3 per cent of all exports) and of industrial goods and materials (C$66 billion) are high, even higher imports lead to significant trade deficits of C$16 billion and C$5 billion respectively in those sectors.[7] The country also has a large trade deficit of C$25 billion in non-automotive consumer goods. These numbers reflect shortcomings in Canada's manufacturing sector. The country's continuing and growing deficit in Information and Communications Technology products is also noteworthy for this study.

Canada's economy is strongly linked to and dependent on the US economy: in 2003, 83 per cent of its exports went to the USA, and 70 per cent of its imports came from the USA.[8] Other major exports destinations are Japan (2.4 per cent in 2003), the European Union (6.0 per cent), and other OECD countries (3.2 per cent). The share of imports from those countries or groups of countries was 3.1 per cent, 10.1 per cent and 5.7 per cent respectively in 2003. Canada's economic activity is therefore very much dependent on the US economy.

Among knowledge industries, two sectors are of special interest to this study: information technologies and biotechnology. Over the past 25 years, Canada's Information and Communications Technologies sector has grown at a fast pace.[9] In 2000, it contributed C$52 billion to the economy (representing 6.6 per cent of the total economy), up from C$36 billion in 1998 and C$44 billion in 1999, at an average annual rate of growth of 11.5 per cent since 1993 compared with 3.5 per cent for the whole economy.[10] The ICT sector has a knowledge-intensive workforce; 50 per cent of its employees have a university degree compared with 19 per cent for the general population, In 1998, it employed 512 000 persons (3.6 per cent of all Canadian workers), 21 per cent being engaged in ICT manufacturing, 37 per cent in software and computer services, 22.3 per cent in communications services, and 20 per cent in ICT wholesaling. Within ICT industries, computer equipment and electronics parts and components (semiconductors) have been the fastest growing production activity, and software and computer services have been the fastest growing service activity. The ICT sector is R&D-intensive: in 2003, at C$5.0 billion, its R&D expenditures represented 41 per cent of total private sector R&D in the country (down from 49 per cent in 2000).[11] About 75 per cent of ICT sector production is exported, mostly to the USA (84 per cent of all exports of ICT goods and services) and to a lesser extent to the European Union (8.3 per cent). However, in spite of the rapid growth of the sector and of the high level of exports (C$39.4 billion in 2000), Canada has a growing deficit in ICT products which reached C$20.9 billion in 2000, up from C$12 billion in 1993. This is due in part to the consistently high level of importation of semiconductors and of electronics parts and components.[12]

The Canadian biotechnology sector is not as well developed as the ICT sector, but it is currently growing very rapidly. In 1997, it had over C$1 billion in product and services sales (40 per cent exported), and was employing about 10 000 people in 282 firms, 25 per cent of which are publicly traded.[13] The majority of those firms are small (72 per cent have less than 50 employees) and fast growing. Close to half of those firms are in the health care sector, followed by agriculture and the environmental sector. Quebec has the largest number of biotechnology firms, followed by Ontario and British Columbia. In 1997, the biotechnology industry spent C$585 million in R&D, 58 per cent in the health care sector, and 37 per cent in agriculture. In 2002, pharmaceutical and medicine R&D alone reached C$1.6 billion, reflecting the very high rate of development of the sector.[14]

History of the science and technology system

In the 1800s and early 1900s, governments were the main performers of scientific activities in Canada. Those activities were directly related to the country's geography and natural resources: creation of the Geological Survey of Canada in 1842 to discover economic minerals and which led to the creation of the federal department of mines in 1907 and bureaus of mines in most provinces; creation of the Toronto Magnetic Observatory in 1840 and Ottawa's Dominion Observatory in 1905; founding of the Central Experimental Farm in Ottawa in 1886 for agricultural research, so important to Canada's 19th century economy; fisheries studies in the Gulf of St Lawrence in the mid-19th century and creation of the St Andrew's Marine Biological Station in 1899, the first of many other such research stations (Canadian Encyclopedia, 1988).

The government's most important initiative in science was the creation in 1916 of the Honorary Advisory Council for Scientific and Industrial Research, which soon became the National Research Council (NRC). The NRC was initially created to encourage industrial research and development in industry and universities with research grants, scholarships and fellowships. It obtained its own laboratories in 1932 and developed solid basic and applied research programs in many fields, some of direct strategic importance during World War II, others of scientific and/or direct economic importance, from nuclear energy, aeronautics and radar to energy, material science, astronomy and chemistry. Provincial governments created similar organizations, such as the Alberta Research Council (1919) and the Ontario Research Foundation (1928). With its C$700 million annual budget, its research laboratories and research institutes, its support programs for industrial research (the Industrial Research Assistance Program, IRAP, in particular), the NRC is still one of the most important elements of Canada's System of Innovation.

A number of government agencies and research support programs were created after World War II, such as the Defence Research Board, the Communication Research Centre, several federal departments supportive of scientific research, and crown corporations[15] engaged in research such as the Atomic Energy of Canada (1952).

In 1969, the federal government created the Medical Research Council (MRC)[16] to promote basic, applied and clinical research in health sciences. The Natural Science and Engineering Research Council (NSERC) and the Social Sciences and Humanities Research Council (SSHRC) were created in 1978 to promote academic research in natural sciences and engineering and in social sciences and humanities. Canada now has a number of excellent universities with research activities supported by government programs and increasingly financed by industry. Research performed in institutions of higher education currently account for a major share of the country's research effort (21 per cent of all R&D performed in the country in 2000, down from a high of 25.6 per cent in 1995).

Science and technology indicators

At 1.92 per cent of GDP in 2001, Canada's gross domestic expenditures on R&D (GERD) trail the OECD average of 2.28 per cent, below the United States (2.74 per cent), but far above Mexico (0.39 per cent).[17] It should be noted that Canada's GERD ratio has increased rapidly and relatively regularly from a low of 1.05 per cent in 1976 (down from 1.25 per cent in 1971) to 1.67 per cent in 1996 and 1.92 per cent in 2001. In comparison, France, Germany, the UK and Italy's GERD ratios have remained relatively constant or have even decreased slightly since 1990, the US ratio has bounced back to its 1990 level after some weakness in the mid-1990s, and the ratios of Japan, Sweden and Finland have continued to increase. In per capita terms, gross R&D expenditures stood at PPP$596 in 2001, compared with PPP$963 in the USA and PPP$36 in Mexico (OECD 2004c, Table 4).[18]

Table 3.1 Canada, science and technology statistics (I)

Gross Domestic Expenditures on R&D (GERD) by funding and performing sectors

	1971	1980	1990	2003
GERD, $ billion, current dollars	1287	3529	10 261	23 293
GERD/GDP (%)	1.25	1.08	1.51	1.91
Distribution per funding sector (%)				
Federal government	44.6	32.6	27.9	19.3
Provincial governments	5.9	7.1	6.2	5.5
Industry	27.0	39.4	38.6	47.5
Higher education	17.6	15.8	15.8	16.5
Non-profit sector	3.0	2.3	2.3	3.0
Foreign sources	1.9	2.9	9.3	8.1
Distribution per performing sector (%)				
Federal government	28.6	20.8	16.1	9.6
Provincial governments	3.3	4.0	3.0	1.4
Industry	33.4	44.5	50.4	53.0
Higher education	33.9	29.9	29.6	35.7
Non-profit sector	0.8	0.9	1.0	0.3

Source: Statistics Canada, Science Statistics, Cat. 88001-XIB and 88001-X1E, v. 28(12), 24 (6) and 18(5), Dec 2004, Dec 2000, and Oct. 1994.

Noteworthy in Canada is the distribution of the R&D activity in the economy: in 2003, industry performed 53 per cent of the R&D carried out in the country, higher education institutions (which includes research hospitals) 35.7 per cent and the federal and provincial governments 9.6 per cent (Table 3.1). This is in sharp contrast to 1971 when industry, higher education and governments were each performing about a third of the national R&D. In 2003, about 47.5 per cent of the funds used for R&D came from industry, 8.1 per cent from foreign sources (mostly foreign businesses), 16.5 per cent from higher education's own funds, and 24.8 per cent from governments (19.3 per cent federal[19], 5.5 per cent provincial). Again, this is in sharp contrast to 1971 when over half of the funds came from the federal and provincial governments, the low level of industrial research being blamed on Canada's natural resource orientation and the level of foreign ownership of its productive sectors, as the important research activities were carried out in the firms' mother country.

In the past 30 years, the development of a few strong 'home grown' knowledge industries (telecommunications in particular) and solid government incentives have led to significant increases in industrial research funding and activities. In 2003, it is estimated that 16 per cent of the country's industrial research activities were in the telecommunication equipment sector (down from 23 per cent in 2000), 7 per cent in aircrafts and parts, 9.5 per cent in pharmaceuticals and medicine, 15 per cent in computer systems design and engineering and scientific consulting and services, 5 per cent in wholesale and trade, and 5 per cent in information and cultural industries for the most active R&D sectors.[20] A major portion of the industrial research (77.7 per cent in 2002) is financed by industry itself, with 14.4 per cent financed from foreign sources, and only 2.2 per cent coming from the federal government (Table 3.2).

Industry funding is playing an increasing role in university research: 'since the mid-1980's, the private sector (industry, private non-profit organizations, etc) has been the fastest growing source of funds for higher education research' (Doutriaux, 2000, p. 96). In 2002–2003, about 24.5 per cent of the research performed in higher education was financed by the federal government, 11.1 per cent by provincial governments, and 8.6 per cent by industry (Table 3.2). Industry is doing an increasing share of its research at universities: 2.25 per cent in 1986–87 and 5.7 per cent in 1996–97 (Doutriaux, 2000, p. 97). In comparison, in the United States, about 6.9 per cent of higher education research was financed by industry in 1995, representing 1.5 per cent of all industry expenditures in the country (NSF and NSB, 1996).

In 2003, there were 156 000 researchers, technicians and support staff employed in R&D activities in Canada (Table 3.2). Of those, 59.1 per cent were in industry, 28.9 per cent in higher education, and 11.4 per cent in federal and provincial laboratories. In per capita terms that represents a 20 per cent

Table 3.2 Canada, science and technology statistics (II)

	Distribution of research funding, by source		Distribution of personnel engaged in R&D, by sector	
	Research performed by industry 2002	Research performed by higher education 2002/03	Researchers, technicians, support staff 2000	Researchers only 2000
Federal government	2.2%	24.5%	9.4%	6.0%
Industry	77.7%	8.6%	59.1%	59.0%
Higher education		46.2%	28.9%	33.2%
Private non-profit sector		8.1%	0.6%	0.2%
Foreign sources	14.4%	1.3%		
Provincial governments and other sources	5.7%	11.1%	2.0%	1.6%
Total expenditures on R&D (C$, million)	12 383	7 428		
Total R&D personnel			156 200	
Researchers only				102 630
Persons per 1000 population			5.04	3.31

Source: Statistics Canada, Science Statistics, Cat. 88001-XIB and 88001-XIE, v 28(10), 28(9), 27(7), 25(6), 25(5), 24(7), Nov. 2004 to Dec 2000.

increase between 1990 and 2003, from 4.18 persons per 1000 population in Canada in 1990 to 5.04 in 2003. When considering the number of researchers alone, Canada, with 6.6 researchers per thousand persons in the labor force in 1999, was below the USA (8.6) and significantly above Mexico (0.6).[21]

Canadian researchers are quite productive, having produced 0.24 publications in science and engineering per researcher in 1999, compared with 0.16 for the USA and 0.13 for Mexico.[22] In addition, globally, in 1999, Canada produced 640 scientific publications per million population, compared with 580 for the US, 400 for the OECD average and 40 for Mexico (OECD 2003b).

In 1999, Canada accounted for about 1.5 per cent of total patent applications to the triadic patent family,[23] compared with 34 per cent for the USA, 32 per cent for the European Union, and 0.03 per cent for Mexico.[24] Per million population, that corresponds to 52 patent applications for the USA, 37 for the EU, 16 for Canada, and less than one for Mexico, Canada being below the OECD average of about 38 patent applications per million population.

Canada has one of the highest rate of internet broadband penetration in the world, with 13 subscribers per 100 inhabitants in 2003 (second only to South Korea), compared with 8 in the USA, 5.5 for OECD and 0.5 for Mexico.[25] Canada has also one of the highest percentages of businesses with access to the internet (90 per cent of firms with over ten employees in 2002) and businesses with their own website (58 per cent in 2002), compared with 10 per cent and 3 per cent respectively for Mexico.[26] Comparable data for the USA is not available.

The supply of venture capital, a key element in the growth and success of many advanced technology firms, is small (0.25 per cent of Canada's GDP between 1998 and 2002) compared with that available in the USA (0.5 per cent of US's GDP) (OECD 2004b, p. 18). As shown in Table 3.3, funds are divided almost equally between early investments and expansion. Early investment does include some seed funding although most of the seed funding still comes from angel investors (informal risk capital) and, to a much lesser extent, government programs. As in the USA, Canadian venture capital investments peaked in 2000. Funds are still available but entrepreneurs must work harder than in the late 1990s to access them. Entrepreneurs are also increasingly relying on other types of early and growth financing.

3.1.2 The National System of Innovation

The institutions and the actors

Scholars have shown that the rate of innovation in a country and its impact on economic activity cannot be explained only in terms of inputs (R&D expenditures, number of researchers) and output (invention disclosures, patents, scientific publications).[27] Key to innovation are the public institutions

Table 3.3 Canada, venture capital invested, by stages (percentages)

	1999	2000	2001	2002	2003
Early	35	46	60	44	51
Acquisition/buyout	6	3	5	3	2
Expansion	53	47	32	50	45
Turnaround	1	1	1	2	1
Other	5	2	2	1	1
Total (C$, total)	2720	6629	4874	2529	1486

Source: Canadian Venture Capital Association, www.cvca.ca, annual press releases.

and the 'private organizations and firms involved in the production, diffusion and commercialization of knowledge, the formal and informal networks which enhance their interactions, the programs and policies that create a supportive framework, and the culturally-driven values that encourage growth and competitiveness.

Central to Canada's National System of Innovation are the thousands of private firms which performed R&D in the country. In 2002, these firms spent C$12.4 billion, and employed 101 000 persons in R&D (4048 with doctorates, 8000 with master's degrees, 51 000 with bachelor's, as well as 26 000 technicians).[28]

Other key actors and institutions include:

- 92 Canadian public and private not-for-profit universities and university colleges,[29] the 12 largest research universities accounting for over 70 per cent of the research expenditures of the group in 1998 (and 24 accounting for 90 per cent of those expenditures) (Doutriaux, 2000, Table 1).
- 128 post-secondary colleges and institutes, many with technology programs.[30]
- Over 120 federal and more than 50 provincial research laboratories, 22 networks of centers of excellence funded by Canada's three federal higher education research granting agencies, Canadian Institute of Health Research, Natural Sciences and Engineering Research Council, and Social Sciences and Humanities Research Council.[31]
- 7 provincial research organizations with C$73 million in research expenditures in 2002.[32]
- The National Research Council of Canada (NRC) with its C$700 million annual budget supporting 17 research institutes and 5 technology centers as well as key programs (described in the next

section), 4000 employees and 1200 guest workers (in 2001–02) reflecting its level of collaboration and partnerships with industry.[33]

- Several industry-led consortia including PRECARN and CANARIE.[34]

The programs

A number of federal programs and organizations support science and technology activities in Canada. Among the most significant are research-granting agencies that support university research:

- The Social Science and Humanities Research Council (SSHRC) for research projects in the social sciences and humanities (2000–2001 budget of C$133.7 million).
- The Natural Sciences and Engineering Research Council (NSERC) for science and engineering (2001–2001 budget of C$550 million), supporting individual research projects in science and engineering as well as 140 industrial research chairs. Several NSERC programs encourage funding partnerships with the private sector. Close to 700 firms are currently involved in such partnerships, about 10 per cent of all the firms carrying out R&D in Canada, twice as many, as ten years ago.
- The Canadian Institutes for Health Research (CIHR), which replaced the Medical Research Council in 2000. It has a 2000–2001 budget of C$402 million and sponsors 13 virtual research institutes.[35]

Several recent initiatives have been introduced to further encourage research and develop linkages between university and industry with particular focus on SMEs:

- The Canadian Foundation for Innovation, created in 1997 to strengthen university research infrastructure (capital investments, equipment), responsible for a budget of C$3.15 billion over several years. Partnership with industry will lead to an expected reinvestment in research of over C$7 billion by 2005.
- The Network of Centers of Excellences (NCE), introduced in 1987, with current funding of over C$70 million (coming in part from the three granting agencies mentioned earlier). In 1999–2000, the Network's 22 Centers of Excellence worked with 563 private companies (which provided C$41 million in additional funding), 138 provincial and federal organizations, 46 hospitals and over 90 universities. One of the objectives of the NCE is to create new bridges between university research and SMEs.
- A new research chairs program, introduced in 2000 with a C$900

million budget to create 2000 chairs at Canadian universities by 2005.

In addition to its research institutes and technology centers, the National Research Council (NRC) has several programs supporting technology partnerships, encouraging university–industry collaboration, and facilitating innovation in SMEs:

- The Industrial Research Assistance program (IRAP)[36] which provides technical assistance and refundable as well as non-refundable financing for R&D projects to small and medium sized firms (up to 500 employees) throughout Canada. This is carried out through a network of 262 Industry Technology Advisors (ITAs) located in universities, research laboratories and economic development organizations throughout the country. In 1999–2000, IRAP reached more than 12 000 SMEs. Its ITAs provided advice on technology and business issues, facilitated contacts and networking with technical colleges and universities, and gave technology development grants to small and medium sized businesses. Projects requiring over C$3 million of support are referred to the IRAP–NRC Technology Partnership Program.
- The Canadian Technology Network (CTN) 'provides pathfinding services to SMEs based on a membership of 1,000 plus innovation-related service providers. In 1999–2000, CTN provided advisory services to 2,300 clients' (IRAP performance report, 1999–2000).
- Precommercialization Assistance Program, established in 1998 jointly with the Federal Department of Industry Canada provides C$30 million annually in repayable loans to Canadian SMEs to help them get their products into the marketplace.
- Canada Institute for Scientific and Technical Information (CISTI), 'North America's largest, most comprehensive provider of scientific, technical and medical information, as well as Canada's leading publisher of scientific journals and books' (NRC 1999–2000 annual report), providing up-to-date information to Canadian researchers as well as a technology watch service to industry (Technology Watch Partnerships).

Similar programs exist at the provincial level. Most provinces have science or research councils providing research grants and facilitating industrial innovation in SMEs. Ontario has a very active Network of Centers of Excellence, established in the late 1970s and precursor to the federal NCE.

3.1.3 Regional Systems of Innovation and the Most Important Technology Innovation Poles in Canada

The best known knowledge regions in Canada are the country's two largest cities, Toronto and Montreal, and three other metropolitan areas, Ottawa, Waterloo-Kitchener and Calgary, which benefited from the telecommunication boom of the late 1990s. Other cities and regions also have a significant technology and innovation base.

Figure 3.1 shows the location of the 11 most important TIPs in Canada, which evolved around local, federal, provincial and private research laboratories, research universities, large technology companies and other anchor organizations. A summary of their socio-economic profiles is presented in Table 3.4, and their universities, science park and incubator facilities are listed in Table 3.5. It must be noted that among those 11 TIPs are Canada's eight largest metropolitan areas (CMAs) as well as the 10th (Waterloo–Kitchener), the 13th (Halifax) and the 17th (Saskatoon), in terms of population size. Levels of education vary widely, Ottawa, Toronto, Halifax,

Figure 3.1 Canada and its main technology innovation poles

Calgary and Vancouver having the highest percentage of population with a university degree. Ottawa and Calgary also have the most science-oriented labor force. Most of these CMAs are the largest urban agglomerations in their respective provinces, accounting for 60 per cent of its province for Winnipeg, 40 to 50 per cent of their provinces for Vancouver, Montreal, Toronto and Halifax, and about 30 per cent each for Edmonton and Calgary. Ottawa, Quebec City and Waterloo–Kitchener are the only exceptions.

Toronto, Montreal, Vancouver, Edmonton, Winnipeg and Halifax have a broad economic base. The other CMAs tend to have a more focused base.

Toronto, Canada's largest city and one of its fastest growing is known for its information technology and telecommunications clusters (the largest in the country), as well as for biomedical and biotechnology activities. Its main competitive advantages are its solid research base, its many public and private research laboratories, and its excellent research universities. Toronto is the capital of the province of Ontario. It is also the main business and financial center of Canada and hosts the head offices of most large Canadian firms.

Montreal, with a broad industrial base, is also known for its solid biopharmaceutical industry and its information technology and aerospace clusters. It is a very research-intensive CMA with excellent research universities and research laboratories, and large multinationals in each of its advanced technology clusters (Merck Frosst, Nortel, Bombardier). The Province of Quebec has also some of the best incentives for R&D and support mechanisms for technology start-ups in Canada. In the 1980s and 1990s, Montreal's economy and its appeal as a high-tech center have suffered from the political situation in the province due to its separatist tendencies.

Vancouver's economy used to be dominated by natural resources-oriented industries. Information technologies, biotechnology, energy, and environmental technologies are now its leading advanced technology sectors. Its local system of innovation includes two excellent universities, a number of research laboratories, the provincial government and its active Science Council. Local quality of life is also a strong competitive advantage, though dampened by a growing cost of living.

The Ottawa CMA (Ottawa–Gatineau) technology orientation is focused on telecommunications, software, photonics, and health sciences. It results from the region's research intensity (especially its large federal research laboratories), the presence of the federal government, Nortel which acted as a large anchor firm, and of other large home grown advanced technology firms (JDS-Uniphase, Mitel, Cognos, Corel).

Edmonton, Alberta's capital city, has a well balanced industrial base with a large petrochemical industry. Its advanced technology clusters include aerospace, biotech, pharmaceuticals, information technology, multimedia, software, electronics and microelectronics. The University of Alberta and the

Table 3.4 Canada's main knowledge poles, basic statistics

Census metropolitan area	Population, thousands, July 2001 (average annual % growth 1996–2001)	Education, % persons 15+ with university degree	Average annual household income, $, 2001	% labor force in natural and applied sciences	Provincial GERD per capita, $, 2002 (industrial R&D per capita, $, 1999)	Provincial VC investment ($millions), Total 1999–2003 (% of Canada's)
Toronto (Ontario)	4868 (2.03%)	18.90	68 400	5.63	Ontario: 811 (461)	Ontario: 8256 (51.7%)
Montreal (Quebec)	3501 (0.62%)	15.07	49 600	5.89	Quebec: 855 (401)	Quebec: 4418 (30.0%)
Vancouver (British Columbia)	2093 (1.82%)	17.23	58 400	5.05	BC: 449 (156)	BC: 1663 (10.3%)
Ottawa–Gatineau (Ontario and Quebec)	1086 (0.90%)	22.60	62 800	9.17	Ontario: 811 (461)	Ontario: 8256 (51.7%)
Edmonton (Alberta)	957 (1.52%)	14.37	57 100	5.35	Alberta: 527 (140)	Alberta: 591 (3.7%)
Calgary (Alberta)	978 (2.96%)	18.53	66 900	8.10	Alberta: 527 (140)	Alberta: 591 (3.7%)
Saskatoon (Saskatchewan)	235 (0.79%)	15.44	50 800	4.53	Saskatchewan: 421 (69)	Saskatchewan: 127 (1.0%)

Winnipeg (Manitoba)	685 (0.15%)	14.72	53900	4.47	Manitoba: 384 (116)	Manitoba: 174 (1.2%)
Waterloo–Kitchener (Ontario)	429 (1.66%)	13.65	59600	4.94	Ontario: 811 (461)	Ontario: 8256 (51.7%)
Quebec (Quebec)	693 (0.28%)	15.77	49200	6.69	Quebec: 866 (401)	Quebec: 4418 (30.0%)
Halifax (Nova Scotia)	361 (1.14%)	18.88	53500	5.02	Nova Scotia: 399 (55)	All Atlantic provinces 284 (2.1%)

Sources: Population, education, annual income: FP Markets, Canadian Demographics 2001, Financial Post Publications, 2001. GERD, Gross Expenditures on Research and Development: Statistics Canada, cat. 88-001 XIE 28(12) Dec 2004. Venture capital investments: www.cvca.ca, Annual Reports, 1999 to 2003, Canadian Venture Capital Association. Adapted from Doutriaux (2003), p. 75, table 6a.

Table 3.5 Canada's main knowledge poles, universities and incubation infrastructure

Census metropolitan area	Major universities (sponsored research, C$ million, FY2003)	Research Parks (number of firms, employment)	Technology incubators (not including VC incubators or multi-tenant buildings)
Toronto (Ontario)	U. Toronto (534) York U. (39) Ryerson Polytechnic(9)	Sheridan Science and Technology Park (18, 2600)	U. Toronto Foundation for Innovation; Food incubator; Toronto Fashion Incubator (new); York Business Opportunity Centre
Montreal (Quebec)	McGill (342) U. Montreal (394) UQAM (60) Concordia (22) INRS (41) Ecole Tech Sup (6)	Laval Science and High-Tech Park (17, 1100) Technopark St. Laurent (4, 230) Technopark Montreal (2 , 150)	CEIM; InnoCentre; Quebec Biotechnology Innovation Center (Laval); NRC Industrial partnership facility
Vancouver (British Columbia)	U. B. C. (349) Simon Fraser (46)	Discovery Parks, 3 locations, at UBC (23, 238), SFU (8, 150), BCIT (9, 1500)	BCIT Venture program; SFU Business Development Centre; Medical Device Development Center
Ottawa–Gatineau (Ontario and Quebec)	U. Ottawa (186) Carleton U. (52)	Ottawa Life Science Technology Park (8, 40)	InnoCentre; Ottawa Life Sciences and Tech.; CRC Innovation Center; NRC Industrial partnership facility; Biotech incubator

Edmonton (Alberta)	U. Alberta (272)	Edmonton Research Park (35, 1200)	Edmonton Advanced Technology Center
Calgary (Alberta)	U. Calgary (165)	University of Calgary Research Park (72, 3000), in 7 buildings	InnoCentre (recent); Technology Enterprise Center (CR&DA)
Saskatoon (Saskatchewan)	U. Saskatchewan (116)	Innovation Place (120, 2000)	
Winnipeg (Manitoba)	U. Manitoba (130)	U. Manitoba SMART Park (2, 50)	NRC Industrial partnership facility; U. Manitoba
Waterloo–Kitchener (Ontario)	Waterloo U. (99) Wilfrid Laurier (11)	Waterloo Technology Park (planned)	
Quebec (Quebec)	U. Laval (287)	Quebec Metro High-Tech Park (102 2500)	CREDEQ
Halifax (Nova Scotia)	Dalhousie U. (79) St. Mary's (3)		

Sources: University research: Re$earch Infosource, http://www.researchinfosource.com/top50.shtml, December 2004. Science Parks: Doutriaux (1998), and regional web sites. Incubators: Interviews and regional development web sites.

Northern Alberta Institute of Technology (NAIT) are two key components of its innovation system, with the Alberta Research Council and other federal and provincial research laboratories.

Calgary is Canada's oil capital and the country's fastest growing city. Its advanced technology clusters in information technology, wireless telecommunications, geomatics and transportation, grew out of research expertise developed to serve the oil industry. Its main competitive advantages are a well educated workforce, excellent university, oil money and low taxes, provincial research laboratories supported by the Alberta Research Council, the country's second highest concentration of large firms' head offices (after Toronto), and several large technology-based firms (TBFs) such as Nortel and Telus.

Thanks to the combined effort of the local, provincial and federal governments, Saskatoon, a small city located in the middle of the Canadian prairies, became a world-level center for agricultural biotechnology and information technology. Its competitive advantages are its concentration of public and private research laboratories in agriculture and biotechnology, its good university, Innovation Place, a proactive research park, and its relatively low costs.

Winnipeg, capital of Manitoba, has a diversified industrial base with branches of many multinational advanced technology firms (AT&T, IBM, Spar Aerospace) and a number of smaller high-tech firms, but with no clearly leading cluster. It has a good research university and several research laboratories (TRLabs (IT), NRC (bio-pharmaceuticals)). It is a low-growth city in a low-growth province.

The Waterloo–Kitchener CMA, originally an active manufacturing area, has emerged as a world class center for information technologies and software. The success of this advanced technology cluster is due to the presence of the University of Waterloo, founded in 1957 as an outward looking university with a strong industrial orientation. The CMA is rich in University of Waterloo spin-off firms, including RIM, the inventor of the 'Blackberry' wireless email solution for mobile professionals.

A quiet government town, Quebec City used to have a limited manufacturing base dominated by wood, paper and food industries. The research activities of its excellent research university and research hospital have now led to the development of small advanced technology clusters in optics, information technologies and telecommunication, and health sciences. Quality of life is one of its competitive advantages, but the region's relatively low research intensity and distance to markets are limiting its growth potential.

Halifax, seat of the Nova Scotia government, has a broad economic base, a good research university, and emerging advanced technology clusters in information technologies, biotech, health sciences, and environmental technologies. Its competitive advantages are a good quality of life and relatively low costs.

Each of these technology innovation poles has at least one very good research university, and most have federal and provincial research laboratories. Toronto, Montreal and Ottawa come far ahead of all the other knowledge regions in terms of R&D activities (total and per capita spending on R&D, number of researchers per capita, federal, provincial and private research laboratories). They also come well ahead of the other TIPs in terms of venture capital investments: from 1999 to 2003, 51.5 per cent of all VC investment in Canada came to Ontario (mainly to Toronto and Ottawa), and 30 per cent to the province of Quebec (mainly to Montreal) (Table 3.4).

Most but not all of those knowledge regions have research parks (Table 3.5), often but not always linked to local universities. As noted in a recent review of Canadian research parks, in the largest metropolitan areas, those research parks tend to be principally providers of serviced land in quality environments, linkages between organizations or with local universities being left solely to each firm's initiative, and business, technical and financial services being available from independent service organizations. In the other large metropolitan areas with active research universities, the most successful parks tend to have been created/managed by regional development authorities with the strong cooperation/involvement of local universities, high-tech regional development being the first objective, and university–industry cooperation coming in second place (Doutriaux, 1998).

Formal technology incubation (Table 3.5) had played only a very limited role in the growth and development of knowledge regions in Canada until recently (Doutriaux 1999). As will be shown in the four regional development case studies which follow, many of the technology incubators created in the 1970s or 1980s have either closed or nurtured only a small percentage of the many start-ups which contributed to regional growth. The situation is changing, with many universities and research laboratories having recently launched or being in the process of launching their incubators.

3.1.4 Summary and Conclusions

Overall, Canada has a good research infrastructure, public and private research laboratories, research universities, and appropriate research programs. Research activities as a percentage of GDP have been increasing over the past 30 years but are still trailing the OECD average. The country's technology policy for many years has been to create incentives to increase private sector research and development activities and to shift responsibility for applied research and its commercialization from the public sector to the private sector. Most public research funding programs involve partnerships with the private sector, leveraging public funds with industry funding. Generous research tax credits coupled with well trained researchers make it also cost-effective for local and multinational firms to perform their R&D in Canada. This trend is likely to continue. A key element of Canada's technology and innovation strategy is the importance given to programs, funding mechanisms and institutional initiatives designed to help SMEs get better access to technology, increase their receptor capacity and develop their ability to use technology as a competitive tool.

Most provinces have at least one TIP, some very large and multi-sector (Toronto and Montreal in particular), others smaller and focused (Saskatoon, Waterloo–Kitchener). Several of these TIPs have evolved naturally from a pre-established industrial base, while others have resulted from a strategic decision by a local institution. Their socio-economic profiles, their scientific bases and the local cultures, are very diverse, as is their level of success. A more detailed analysis, presented in the next section, is needed to identify success factors and trends.

3.2 CANADIAN CASE STUDIES

As noted previously, the objective of this book on knowledge regions in North America is to analyze selected knowledge regions in the USA, Canada and Mexico, and to derive lessons on the approaches, institutions and policies that are appropriate in one national environment and less so in another. Special focus is put on the mechanisms used for nurturing innovative firms and fostering their agglomeration in each region.

The four knowledge CMAs selected to represent the Canadian experience were chosen among those of the 11 knowledge regions described in the previous section with at least one science park or one technology incubator, to illustrate specific conditions:

- Ottawa, as an example of a high R&D-intensive government town which developed an entrepreneurial culture as its advanced technology clusters grew.
- Saskatoon, to illustrate the cooperation of the provincial and federal governments to develop a world class focused cluster in a geographically isolated small city.
- Calgary, as the result of the provincial government decision to develop an advanced technology cluster to diversify the oil-based local economy.

- Quebec City, as an example of the development of advanced technology clusters based on local academic research expertise in a moderately isolated government town.

3.2.1 The Ottawa Area

Seat of the Federal Government of Canada since 1867, Ottawa was a typical government town until the mid-1970s, the government sector and government-related services dominating the local economy. The Ottawa Census Metropolitan Area, comprising the cities of Ottawa (Province of Ontario) and of Gatineau (Province of Quebec), is located 200 kilometers west of Montreal, 400 kilometers north-east of Toronto in what used to be a forest-products sawmilling region. In 1969, the Ottawa region had 430 000 inhabitants, 33 per cent of its workforce (57 490 persons) worked for the federal government, and it had a very limited industrial base. In 2003, the region had over 1.1 million inhabitants, and only 18.1 per cent of its workforce (72 825 people) worked for the federal government. Since the mid-1970s, it had developed a healthy high-technology sector, from an estimated 2000 high-tech jobs in 32 firms in 1950, to 8200 jobs in 160 firms in 1975, to 50 000 jobs in 900 firms in 1998 (with total sales of over C$3.4 billion), and 79 000 jobs in 1050 firms in 2000. High-tech employment had passed federal government employment in mid-2000, the result of government downsizing and of the high growth of high-technology but consequently fell back slightly. In 2003, high-tech employment had retreated to 64 000 jobs but the number of TBFs in Ottawa was back at over 1400, with 12 per cent of firms and 18 per cent of employment in telecommunications, 25 per cent and 21 per cent respectively in software, 3.3 per cent of firms and 10.6 per cent of employment in photonics (a sector heavily fueled by venture capital around the year 2000), 7 per cent of firms and 15 per cent of employment in microelectronics, and 3.5 per cent of firms and 3.4 per cent of employment in life sciences.[37] In mid-2004, regional high-tech employment increased to 67 000, and the number of TBFs continued to increase to close to 1600 (Figure 3.2).

Regional technology infrastructure

High-technology benefited from the research activities of government laboratories (National Research Council (NRC, founded in 1916), the Defence Research Establishment (late 1940s), a precursor of the Communication Research Centre (CRC, created in 1969) in particular), and of the presence in town of a few small electrical and cable companies as well as of Northern Electric (created in 1914). Northern Electric evolved into Northern Telecom (1960) and then, after its merger with Bell Northern Research, into Nortel, Ottawa's current largest private sector employer (17 000 jobs in the region in

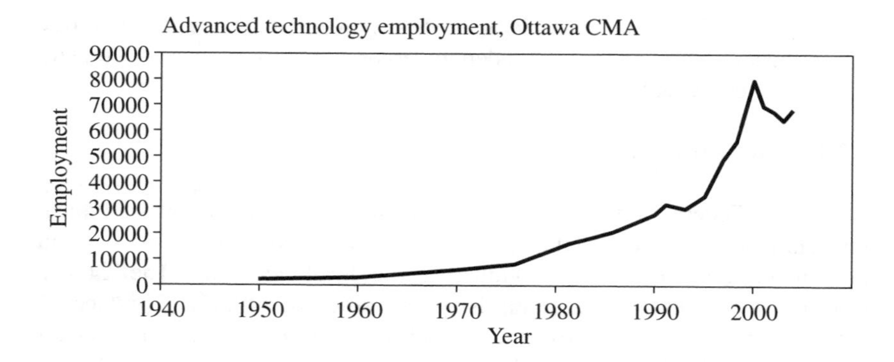

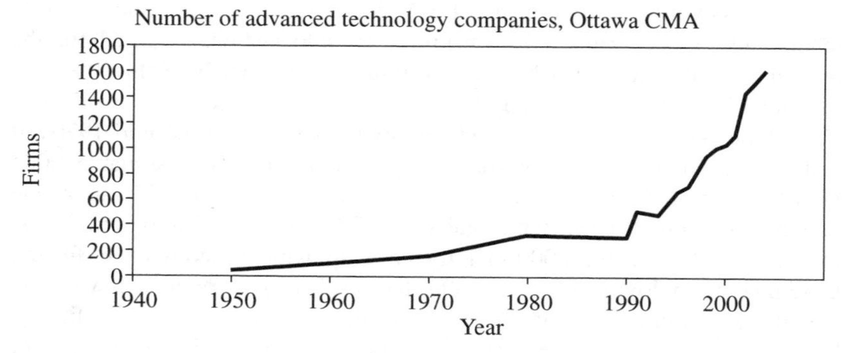

Source: OCRI

Figure 3.2 Ottawa CMA advanced technology employment and companies

mid-2000), and which has directly or indirectly spun out over 120 firms over the years. The first real high-tech firm in the region, Computing Devices, was created in 1948 (NRC's first spin-off), and has itself spun out over 30 companies.

The real starting point of local high-tech development can be traced to the 1971 failure of Microsystems International, a two-year-old entrepreneurial subsidiary of Northern Telecom. Its bankruptcy released a number of engineers and scientists with entrepreneurial aspirations who went on to launch a number of start-ups. Two of those technology entrepreneurs were especially successful, Terry Matthews and Michael Cowpland. They created Mitel in 1973 and sold it to British Telecom in the late 1970s. Terry Matthews then created Newbridge (sold in 2000 to Alcatel) which then incubated and spun off a number of other successful firms; Michael Cowpland created Corel, which soon became Canada's largest software company. They became role models in the community, contributing to the development of a solid entrepreneurial spirit. This nascent entrepreneurial movement benefited from the presence in town of Noranda Enterprises Ltd, the venture capital fund of a large Canadian natural resource company that invested in a number of local start-ups in the mid-1970s. Local development then rode the microelectronic/ telecommunications technology wave, reaching a critical mass in the mid-1980s, the local high-tech cluster being large enough to support the business and technical infrastructure needed for sustained growth. The creation of the Ottawa Carleton Research Institute (OCRI)[38] in 1983, a non-profit, self-financing consortium of local universities and colleges, private firms and governments, contributed very significantly to the development of the region: from the start, OCRI set out to develop strong networks between firms, universities, colleges and governments, encouraging partnerships and cooperation, supporting the creation of research chairs at local universities, offering training seminars and networking opportunities. At first, the two local universities (University of Ottawa, Carleton University) and post-secondary colleges (Algonquin College and la Cité Collégiale) were only indirectly involved in the high-tech development of the region, and then mostly as suppliers of trained manpower. Over time, they developed their research linkages with local firms. They are now active contributors to local development, their research being often done in partnership with local firms, and their academic programs in science and engineering being finely tuned to better meet the needs of the industry. In 1997–98, 51 per cent of the research carried out at the University of Ottawa and 31 per cent of the research at Carleton University was financed directly by industry and private foundations.[39]

Ottawa has long been known as a 'home grown region', a large number of its 1600 firms (2004) being local start-ups: Mitel, Newbridge (acquired by Alcatel in early 2000), Mosaid, JDS-Fitel (now JDS Uniphase, the world's largest fiber optics equipment maker in the early 2000s), Cognos and Corel. Many multinationals are now established locally: Nokia, Ericsson, Siemens and Lucent Technologies. Until the mid-1980s, local venture capital was limited. In the early 1990s, 'angel' funding started to be available, but it was not until the mid-1990s that significant amounts of venture capital became available in the region. VC firms invested an average of C$100 to 250 million annually in local firms in the late 1990s and reached C$1.2 billion in 2000. It has since decreased following the North American trend, totaling about C$2.1 billion for 2001–2004 combined. In the early 2000s, there were several local VC firms (such as Capital Alliance Ventures, Celtic House) as well as offices for a number of US firms (such as Mohr Davidson, Greylock, Kodiak Venture Partners). In 2001, OCRI absorbed the region's Economic Development Organisation, combining its expertise in high-technology with OED's expertise in economic development and planning. The 'new' OCRI, 'rallying point to advance Ottawa's globally competitive economy and superior quality of life',[40] continues to foster the development of the local knowledge economy. Growing and retaining local high-tech start-ups has become one of its priorities, in addition to its current activities designed to enhance local networking and collaboration between all segments of the economy and of society, and to support initiatives that bring together the ideas and people that make Canada's advanced technology industry thrive.

High-technology has changed local attitudes, from those of a sleepy government town to the dynamism of a Silicon Valley style town, without the costs and the traffic jams. High-technology is very visible in the local media and also in local social activities (high-tech sponsorship of charitable fundraising events, of sports facilities, of local artistic and cultural events).

One special characteristic of the Ottawa region is its research concentration and intensity, by far the highest in Canada: In 1998, total local private sector R&D expenditures were estimated at C$1.7 billion[41] (89.5 per cent in telecommunications, 7.5 per cent in software, 2.2 per cent in hardware, and 0.8 per cent in life sciences). Public sector R&D expenditures (by the federal government) were C$805 million.[42] Including university research, that translated into about C$2700 of R&D expenditure per capita, far above the country's average which was then about C$500 per person, and more than twice as high as what was then the per capita R&D expenditures in Montreal and Toronto. Since 2000, with the telecommunication industry slowdown, local R&D by private firms has decreased, but the region's R&D intensity (expenditures per capita) is still higher than any other region in the country.[43] Extensive networking activities between government researchers, private firms and universities also continue to make a positive impact on research activities and its commercialization.

Science park and incubators

Formal technology incubation activities in the region are relatively recent. Although a significant number of the local TBFs are 'home grown', they were started by engineers and scientists from larger high-tech companies and government research labs as independent ventures or, in a smaller number of cases, as subsidiaries. Only a limited number of start-ups actually came from the universities. Until the early 1990s, technology incubation remained a private matter between the entrepreneurs and their private technical and business networks. In 1992, the regional government created a very successful entrepreneurship centre[44] which operated as a virtual incubator, offering advice, support and training to thousands of local would-be entrepreneurs and facilitating access to networks of mentors and private investors.

Formal efforts at technology incubation activities started also in 1992 with the creation of the Ottawa Life Sciences Technology Park (OLSTP, a 40000 square foot multi-tenant building (8 firms, 40 employees) on 9 hectares of land in the park located next to the city's main research hospitals and the medical school and faculty of Health Sciences of the University of Ottawa).[45] The objective of the Ottawa Life Science Council, a consortium of local firms, research laboratories, governments and the University of Ottawa, was to leverage local health research expertise (medical school, research hospitals, NRC research institutes in health and biology) to develop a local biotech industry (with estimated 1995 sales of C$300 million).

This was followed in 1994 with the opening of the CRC Innovation Centre, and in 1998, of NRC's Industry Partnership Facility. The CRC Innovation Centre provides space (10 000 square feet of rentable space in 1999, very flexible), and direct access to CRC technologies, facilities and expertise. To be admitted, start-ups and research units of existing firms must be sponsored by a CRC researcher and demonstrate the expected advantages of close cooperation. Small business counseling, an on-site NRC Industrial Technology Advisor, direct access to laboratories, researchers and technical information, are some of the benefits derived from incubating at CRC. In mid-1999, this incubator had 11 tenants with about 80 employees; over 20 tenants had graduated, 15 still alive and growing well. Rent at or above market rate, is not an issue, the benefits of co-location with CRC being more important. Normal maximum tenancy is for two years, with an informal review every four months. The main advantages of this very successful incubator (from a technology transfer point of view) are the research mentorship provided by the CRC 'sponsor', easy access to test beds, and the speed and ease of entry into the incubator for qualified candidates (limited bureaucracy, responsive process). The NRC Industry Partnership Facility (IPF, 35 000 square feet of space; 14 000 available for rent), created in 1998 and collocated with two of NRC's research institutes (similar facilities have now been built in Montreal and Calgary) had objectives similar to CRC's: collocation of researchers and industry (start-ups or research units of existing firms) to facilitate technology development and transfer. Space, common areas and basic services, business counseling, networking opportunities, direct access to researchers, sophisticated equipment, and technology counseling are some of the benefits received by tenants who, as at CRC, pay above market rental prices (rental rates increase proportionately to gross sales). Normal maximum tenancy is for three years, but actual lengths of stay vary depending on need.

Two large private communications firms, Nortel and Newbridge, also launched formal technology incubation programs in the 1990s: the Nortel business incubation program was initiated in 1996 on the venture capital model to further develop internal inventions with commercial potential and provide an 'in-house' alternative to employees wanting to become entrepreneurs.[46] Between 1992 and its purchase by Alcatel in 1999 for about

US$7 billion, Newbridge has developed a very successful 'affiliates program', nurturing start-ups with internal or external technologies complementary to its own or having the potential to expand its markets. Affiliates were financed personally by Newbridge founder Terry Matthews, and also by Newbridge, and were provided with space and full access to Newbridge's marketing, legal and technical resources. This corporate incubation program proved to be extremely successful,[47] a great part of the credit going to Terry Matthews and his skills at spotting hot technologies. The late 1990s have seen the develop-ment in Ottawa of a number of private venture capital incubators owned by local 'angels', former engineers and scientists who became wealthy with their stock options when the firms they were working for became public in the mid or late 1990s. Others are supported by external VC funds: Innocentre (main office in Montreal; working closely with the NRC in Ottawa), Raza Foundries (main office in California), StartingStartups, Itemus, Momentous.ca, OnX Incorporated, Reid Eddison Inc., Venbridge Inc., Venture Coaches, to name a few. Thanks to the provincial government, City of Ottawa, and private funding, a new incubator, the Greenbelt Research Farm Facility (18 000 square feet) was opened in 2001 by the Ottawa Biotechnology Incubation Centre (OBIC) on land owned by the Federal Government. In early 2002, another incubator (26 000 square feet was being built in the Ottawa Life Sciences Technology Park, and plans for two other incubators (26 000 and 80 000 square feet) were being made to meet the expected demand. An estimated 11 000 people were employed in biotechnology and health sciences in Ottawa in 2002.[48]

TBFs surveyed

Interviews were conducted during the summer of 2000 with the directors of Ottawa's three formal technology incubation facilities, the CRC Innovation Centre (CRC-IC), the NRC Industry Partnership Facility (NRC-IPF), and the multi-tenant facility of the Ottawa Life Science Technology Park (OLSTP). A survey of their tenants was organized to evaluate the impact of incubation on their activities. A copy of the questionnaire used for the survey was mailed to the 22 firms listed as tenants in the incubation facilities or hand delivered in their mail boxes (Table 3.6). Up to three follow-up telephone calls were made to arrange for in-person interviews. Three firms formally declined to participate, citing confidentiality issues. Seven of the eight firms which did participate in the study completed the questionnaire on their own, lack of time being blamed for their inability to participate in in-person interviews.

The CRC-IC and the NRC-IPF objective is to facilitate technology development and technology transfer from their laboratories to industry through collocation of researchers and industry (start-ups or research units of existing firms). To be admitted, potential tenants must clearly show the expected synergy from collocation in the laboratory and be sponsored by a

Table 3.6 Ottawa CMA: TBFs in incubators interviewed

Incubation facility	Number of TBFs in facility (listed as tenant in 2000)	Number that formally declined to participate in study	Number of answers (response) rate)
OLSTP	6	1	3 (50%)
NRC-IPF	7	2	1 (14%)
CRC-IC	9	0	4 (44%)
Total	22	3	8 (36%)

local researcher. Admission gives them access to all the services available at CRC or NRC (laboratories, equipment, library and information services). The OLSTP operates as a heath science multi-tenant facility located close to Ottawa's major hospitals and to the medical school and health sciences faculty of the University of Ottawa. It has a clear sectoral focus but does not require prior joint research plans or partnerships between tenants or with local researchers. It does organize regular networking opportunities, including the well-attended 'luncheon at the park' conference series. None of the three incubators provide formal business counseling to their tenants. Such services used to be available at the CRC but were discontinued as excellent entrepreneurial support is available throughout the region. Instead, start-up entrepreneurs are referred to the Entrepreneurship Centre (a successful regional small business counseling center), to Innocentre (located in the OLSTP), or put in contact with the many high-tech support groups, angels, mentors and VCs in town.

The eight TBFs that participated in the survey are fairly typical technology-based incubating start-ups in photonics (three firms), information technologies and software (four firms), life-sciences (one firm); median age two years (a few months to 12 years); median employment seven employees (one to 16 employees); three firms with no sales yet; 25 per cent to 100 per cent local sales (IT and software firms having a local orientation). The firms reported one to five product or service innovations (median two), product development being mostly market-driven (30 per cent to 100 per cent market pull) and performed mostly internally. Only two firms had patents (one and two patents), the other firms relying on informal IP protection. All the firms reported having good links with researchers in their respective laboratories (NRC, CRC) and with the universities (for the OLSTP tenant), and occasional links with government, local professional and technical associations, local business groups, and law firms. Biggest advantages of incubation were security (controlled access facility), easy access to R&D personnel and facilities, flexible rental space, and noted by only a small number of respondents, image and business support. Responses clearly indicated that the main benefit resulting from these start-ups was physical closeness to local public laboratories, research hospitals and universities. The Ottawa region, with its mix of large and small technology firms, government, public and private laboratories, universities and colleges, strong business networks, experienced law firms, and informal and formal venture capital, was considered to be the primary comparative advantage for firms over competing firms from other regions, before even their incubator. One local disadvantage in the year 2000 was the level of local competition for scarce manpower.

Conclusion

Overall, formal technology incubation has played only a limited role in Ottawa's development as a high-tech cluster. Six periods/stages can be distinguished in Ottawa's path to development, specific actions designed to support the development of a technology cluster starting only with stage three (Brouard et al, 2004, p. 62).

- Stage one (until 1970): The high level of government R&D and government procurements provided an environment conducive to start-ups: unsolicited proposals for applied research and development, subcontracts with the government, access to government procurements (one of the most important success factors for Canadian start-ups in the 1970s and early 1980s (Doutriaux, 1992)). Creating a company was also relatively simple (few regulations, easy incorporation) even in the absence of local investment capital.
- Stage two (1970s): Growth of BNR/Northern Telecom/Nortel, a large research and manufacturing telecommunications company which developed its research laboratories in Ottawa to be close to the government and its research laboratories, and acted as a large private-sector anchor organization, making the region more attractive to other private-sector ventures; small number of star-ups, some of which became very successful (Mitel in particular).
- Stage three (1980s, early 1990s): slow growth of the local telecommunication, microelectronics, software sectors, 'home-grown' start-ups following in the steps of solid local role models (Mitel, Mosaid). Creation of OCRI, a private sector–universities–government consortium intent on developing successful business–university–government partnerships, creating occasions for active networking activities (Monthly Technology Executive Breakfasts, monthly Zone 5 meetings primarily aimed at technology marketing and so on), develop-

ment of better communication links (direct flights to major US cities), development of high-tech support services (precision machining, precision mouldings and so on); growth of angel networks (wealthy local entrepreneurs).

- Stage four (1994–1997): international recognition and high regional growth. Development of local venture capital; entry of US VCs following the highly visible purchases of local start-ups by large multinationals and some very successful IPOs; arrival in the region of external firms (Cisco, Compaq, Nokia and others).
- Stage five (1998–2000): 'second generation VCs': wealthy local entrepreneurs create their own VC funds and start nurturing local firms; creation of public and private incubators, high rate of start-ups; growing interest for medical and health sciences (partnership of universities, NRC, hospitals).
- Stage six (post 2000 and current status): relative regional slowdown with large layoffs by the largest corporations; high rate of start-ups by laid-off employees bringing the number of TBFs to an all-time high in 2004; creation of biotech incubators, health sciences and biotechnology being seen as potentially the next technology wave. After decreasing by 20 per cent between 2000 and 2003, high-tech employment has started to grow again in 2004, a positive sign for the region.

The Ottawa CMA has all the elements needed for long-term sustainability and there is no doubt that the region will continue its recovery from the technology meltdown of the early 2000s. There is a solid supply of knowledge workers, a culture that is very supportive of innovation and entrepreneurship, very active business and technical networks facilitating communications and exchanges between engineers, scientist and business people, committed organizations, such as OCRI, working jointly with the regional government, local universities, large corporations to promote research and technology commercialization locally, and excellent quality of life. Start-ups must work harder than during the dot.com boom of the late 1990s to get funding but angel investments and venture capital funding is still available (about C$200 million of venture capital in 19 deals in 2004).[49] The lesson that was learned is the need to diversify the local high-tech base, overly dependent on telecommunications and photonics in the late 1990s, hence the current efforts to develop health sciences and biotechnology as a complement to the existing telecommunications, software, and microelectronics clusters.

3.2.2 The Quebec City Metropolitan Area

Capital of the Province of Quebec, and its second largest urban area, Quebec

City and its metropolitan area had 693 000 inhabitants in 2003. One of the oldest towns in Canada, it is located on the north shore of the St. Lawrence river at the eastern end of the Quebec–Windsor corridor, Canada's most populated and economically active region. Quebec City is relatively isolated as it lies 250 kilometers from Montreal, 450 kilometers from Ottawa and 800 kilometers from Toronto. Quebec tends to be a stable, traditional government city. It was barely affected by the 2000 dot.com meltdown. It has one of the lowest birth rates in the Province of Quebec and a low rate of population growth, 0.3 per cent annually on average between 1996 to 2001 (Table 3.4), compared with 0.5 per cent for the province, 0.6 per cent for Montreal, and 2.03 per cent for Toronto. The local level of education is average, 15.8 per cent of the population over 15 having a university degree, compared with 15.1 per cent in Montreal, 18.9 per cent in Toronto, and 22.6 per cent in Ottawa, Canada's most educated city (Table 3.4).

The economic development of the Quebec City area started in the 17th century as an important port of entry and exit for ocean-going sail boats and a transfer point between domestic and foreign trade. Its location made it an important administrative and political center in the early years of the development of what is now Canada. In the late 19th and early 20th centuries, the development of steam-powered ocean-going vessels able to go directly to Montreal, delays in establishing a rail link to the city[50] and the general shift of business westward, made it lose its economic dominance to Montreal. In spite of significant efforts made over the past century to develop manufacturing activities and diversify its economic base, most jobs in the Quebec City metropolitan area are still concentrated in the service sector (public administration, defense, and the service sector (85 per cent of local employment). Only 10 per cent of local employment[51] is related to manufacturing (paper and related products, printing and editing, food processing, metal products, and a large oil refinery).

Regional technology infrastructure

The late 1980s and 1990s have seen the slow development of promising high-tech activities with the creation of several research laboratories and technology-based enterprises in sectors related to local academic research, health sciences and government activities. Between 1991 and 2001, employment in science and technology activities has grown from about 4 per cent of total employment to 6 per cent, while public service employment has decreased slightly from 13 per cent to about 12 per cent.

The Quebec region is a moderately research-intensive region. In 2001, R&D expenditures per capita in the region were estimated at C$990, only slightly above the provincial average (C$838) and far below Montreal (C$2044), the leading R&D pole in the province:[52] 56 per cent of Quebec

City's R&D was carried out in the academic sector (mainly Université Laval and its affiliated hospitals), 27 per cent by industry and 17 per cent by the provincial and federal governments, compared with 29 per cent academic, 69 per cent industry and 2 per cent governments in Montreal. In spite of that moderate level of R&D activities, the region has been able to develop a small, dynamic, high-tech sector based on a successful niche strategy. According to the local economic development organization (Société de promotion économique du Québec métropolitain),[53] over 6000 researchers and support personnel are currently working in over 100 research centers in the region. Among the best known are the Institut National d'Optique (optics and photonics), the Val Cartier National Defence Research Centre, the research center of the Hospital of Université Laval (health sciences), Université Laval's Geomatics Institute and Intermag technologies, a fast prototyping company. Photonics, geomatics, information technologies and health sciences, are the main domains of local high-technology activity. Their development can be traced to the research and expertise at Université Laval and in its affiliated teaching hospital. In 2000, about 4.1 per cent of local employment was in the information and communication technologies sector (about 300 firms), a growing proportion of the local workforce, but still low compared with Ottawa's 13.4 per cent.[54]

Science park and incubators

Formal technology development and incubation activities in the region started in the early 1980s through a joint effort of Université Laval, local government and the local business elite. The first formal technology incubation infrastructure in the region was a large research park located on 335 acres of land not far from the campus of Université Laval in the Quebec city metropolitan area. Planning started in 1983, a collaborative effort of the local government, the governments of the Province of Quebec and of Canada, local businesses and Université Laval. The Parc de Haute Technologie de Québec Métropolitain was formally created in 1988 as a joint venture between the federal government (30 per cent of current operating costs), the provincial government (25 per cent), and the local municipalities. Its objective was to be an agent of regional economic diversification by facilitating and encouraging the implantation or creation and development of public and private research centers and of high-tech enterprises in a setting conducive to research and innovation, public–private–academic networking and partnerships. Entry in the park was reserved to research-intensive organizations (research laboratories and high-tech firms). Residents in the park could build their own facilities or rent space in a privately built and managed multi-tenant building (in early 2000, there were four such multi-tenant buildings, owned and managed by the same real estate firm). Until the late 1990s, the Park was managed by a non-profit corporation, the 'Corporation du park technologique', an initially very proactive organization which organized a large number of support activities (information sessions, conferences, workshops, networking activities and marketing). Linkages with the university were quite strong; more than half of the members of the park's board of directors came from the university and the university technology transfer office had a permanent office in the park. The park developed rapidly, the combination of an excellent working relationship between the various park sponsors (three levels of government, university, business community), a prestigious address, and also probably, the attraction created by a large government research laboratory, the Centre de Recherche Industrielle du Quebec (CRIQ) which was located there at the time of creation and was soon joined by two other large research centers, the Institut National d'Optique (INO, a provincially-funded research center linked to Université Laval) and Forintec (a woods industry research center). Those early park residents were quickly followed by ten other research centers and a number of technology-based firms. By 1994, there were 85 firms and research organizations located in the park, 10 firms being spin-offs from Université Laval and seven from INO. Communications with the university were facilitated by the presence in the park of an office of the Bureau de Valorisation de la Recherche (BVAR), which acted and still acts as 'a door to the university for residents in the park'. Park growth and development then slowed down. In the mid to late 1990s, the park went through a crisis: changing leadership, reduction of organized networking and marketing opportunities, elimination of services to residents, questions about the park's long-term sustainability, and lack of funding (the park was still completely dependent on government subsidies as tenants were not asked to contribute to operating costs, and the governments indicated their plans to phase-out their support). In 1999, the park was privatized, the development and sale of its undeveloped land being expected to finance its operations. The park is now managed by a team of seven and has a large board of directors (26 persons from business, various government entities and the university). Some support services to residents were resumed (business assistance to SMEs, technology assessment, business planning, marketing and export promotion). Three years of bridge financing by the original sponsors was provided to facilitate the transition to private operation (C$250 000 each per year for three years from the federal and provincial governments, C$300 000 from the local municipalities). By 2001, there were over 100 firms and organizations in the park with total employment of over 2500 and still a lot of open space for development. The park does not have a strong sectoral focus, firms being active in domains related to the region's traditional industries (forestry) or to its research base (optoelectronics, information technologies and telecommunications, new materials and nordicity).

The region's first business/technology incubator was created in 1988–89. In 1988, a local community college (offering three-year post-secondary programs) created a business incubator. In 1989, Université Laval decided to join in to provide support to its own technology spin-offs. In partnerhip with the city of Quebec, the Fondation de l'Entrepreneurship, and the Société Québécoise de la Main d'Oeuvre (SQDM), a job creation public organization, this led to the creation of the Centre Régional de Développement d'Entreprises de Québec (CREDEQ), financed by the Quebec and federal governments as well as local businesses. The incubator operated from two locations: two small buildings in town (24 000 and 18 000 square feet) acquired at a very low price from the municipality, and leased space in the Parc Technologique du Québec Métropolitain, in colocation with research centers and high-tech firms, and with easy access to Université Laval's technology transfer office. The objective of CREDEQ was to contribute to regional development by providing space and support services to high-tech start-ups (university and college spin-offs and others).

In 1996, the CREDEQ board of directors of 15 members included two representatives of the university and two of the college. Incubated firms had no specific sectoral orientation. Until 1996, support services, business counseling and business planning were offered free of charge to incubator tenants who also had subsidized rents. Thirty-eight firms were admitted by CREDEQ between 1988 and 1996 for two to three years of incubation. By 1997, 31 had graduated and 28 were still alive (an excellent survival rate), with combined sales of C$29 million and 360 employees. Of the seven firms in the incubator at that time, four were developing new technologies in various domains; the others were using known technologies and working as subcontractors for other firms in the region. Relations with the university in the form of use of the expertise of professors, use of students as manpower, use of university equipment and laboratories varied from firm to firm. The incubator also benefited from the very active Centre d'Entrepreneuriat et de PME of Université Laval (Entrepreneurship and small business centre). Under the leadership of Professor Yvon Gasse, this center has been at the forefront of entrepreneurship education in Canada. It still offers a number of programs at various levels (from courses offered to the general public through the media or at night, to Master's and PhD programs) and, through Entrepreneuriat Laval, offers support services to local entrepreneurs (such as research, seminars and opportunities for networking).

Since 1997, the situation has changed as CREDEQ's main sponsors, the federal and provincial governments, asked the incubator to become self-sufficient and announced that their funding would end in late 2002. In early 2000, CREDEQ started to take stock options in incubating firms and to charge them a C$52 000 fee for 400 hours of business coaching and counseling in addition to the (subsidized) rent paid for the space used. The fee is in fact a repayable loan, to be repaid at a charge of 2 per cent of gross sales after one year of incubation. CREDEQ also charges 4 per cent of the venture capital raised while in incubation. Three incubatees left CREDEQ in mid-2000 when the new fee was introduced, leaving only three firms in incubation (two tenants, one outside). Between 1996 and 2000, the number of CREDEQ employees (administration, business support) had decreased from eight to two. In mid-2000, the CREDEQ realized that it could not raise the funds needed for its operations. It was planning to change its mode of operation, from a 'bricks and mortar' incubator to an incubator without walls, providing business and counseling services to start-ups irrespective of physical location.

Between 1988 and 2000, CREDEQ has incubated 52 firms. In 1998, after ten years of operations, the total sales of the survivors totaled C$52 million.[55] A directory published by CREDEQ in January 2000 lists 41 companies with total employment of 490 persons (1 to 150 employees). Sectors of activity are very diverse and include software (25 per cent of the firms), information technologies, biotechnology, optics, mechanical engineering, services and other types of activity.

Part of the reason for the reduced demand for the services offered by CREDEQ as a brick and mortar incubator is probably due to its lack of sectoral focus, new tenants not getting the benefit of a colocation with firms with similar or complementary experiences and networks. In the words of one of the entrepreneurs who was interviewed for this research, 'I did not want to be isolated in the incubator; I wanted to be directly in the Park where the action is' (he was referring to the two buildings CREDEQ has in town as the incubator had stopped leasing space in the park). Other reasons for the reduced demand for the type of service offered by CREDEQ are the changing needs of start-ups looking for space in facilities with a better defined sectoral focus, and the availability of specialized counseling and support services for start-ups from many external sources, in addition to the internet:

- The Centres Locaux de Développement (CLDs), a program of local development centers created in 1997 and sponsored by the province, offer start-up support services and business counseling in many different locations; CREDEQ has very good relations with 10 local CLDs.
- Specialized incubators with a clear sectoral focus have recently been created in the town: a fashion incubator, and a multi-media and information technology incubator located in a centrally located former industrial building as part of the city's core revitalization.
- Entrepreneurial counseling and/or coaching is readily available from Entrepreneuriat Laval, a small-business support activity offered by Université Laval.

- Private or public venture capital funds also provide the coaching and mentoring needed by the high-tech start-ups in which they invest.

Start-ups in Quebec City do not have access to the same amount of venture capital as their counterparts in Montreal, Ottawa or Toronto. This shortcoming is partially corrected by Innovatech, a venture capital fund set-up by the provincial government to support high-tech start ups. Innovatech has invested over C$62 million in over 90 local start-ups in recent years,[56] providing at the same time counseling, networking and mentoring services.

TBFs surveyed

Two surveys of CREDEQ tenant firms were performed, one in 1996 and one in 2000, to evaluate their incubation experience.

In 1996, before the start of this project, a short two-page questionnaire (15 minutes average completion time) had been sent to the seven tenants who were then in the incubator. The entrepreneurs were encouraged to return their questionnaire using a toll-free fax number to the University of Ottawa. A total of five questionnaires were returned, a 70 per cent response rate. Two firms were in software development, the others in light manufacturing in three different domains. All firms were users of technology rather than technology developers. At time of entry in the incubator, they were between 0 and 4 years' old without sales or with sales up to C$35 000. At interview time, median age was two years and sales between C$15 000 and C$600 000. All firms grew during their two to three-year incubation, but only one experienced really high growth (over 30 per cent per year). Reasons for entry were purely business-related (access to business assistance and networks, possibility to make contacts with other firms, shared/cheaper space and office services), with no special interest in university linkages or access to technology. Only one firm reported access to technology and expertise and the possibility of joint research with academics as a major benefit.

In 2000, a French version of the questionnaire designed for the 'Monarca project' was sent to the 41 entrepreneurs listed in CREDEQ's directory of tenants and graduates.[57] In spite of follow-up telephone calls, only six entrepreneurs (two firms still in incubation and four graduates) responded to the survey, a low 15 per cent response rate. The length and complexity of the questionnaire and the fact that in most cases, CREDEQ incubation took place several years before the survey, partially explains this very low response rate. Because of the small sample size and the fact that entrepreneurs did not complete all the sections of the questionnaire, the descriptive analysis which follows cannot be considered representative of CREDEQ graduates. Two firms were in information technologies, two in medical products, one in industrial manufacturing, and one in a high-tech niche consumer product.

Three firms have some manufacturing activities. Responding firms were relatively young (one to eight years old) and small (one to nine employees, with mean sales of C$200 000). The respondents listed from none to three product innovations with a median of two, and no managerial innovations. Product innovations were driven by market pull (60 per cent) rather than technology push (40 per cent). Technology/product development was done mostly in-house (80 per cent), little outside (20 per cent). None of the firms had patents; one firm noted that it did not want to attract visibility by disclosing what it considered to be its industrial secret. All the firms noted the importance of R&D, even in the most applied cases. The most important services received from the incubator were of a financial nature (assistance in accounting, help in obtaining access to financial sources, angels and investors), followed by business services (business planning, marketing, legal assistance and referrals), and training. Two firms noted the increased credibility they gained when admitted by CREDEQ, making it easier to get banking loans. Several noted the benefits from networking with other incubatees and the help they received from CREDEQ when trying to access government services. Proximity to the airport, and the presence of the Parc Technologique were advantages noted by a few entrepreneurs; disadvantages included the small size of local markets, the relative isolation of Quebec City (far from the large urban centers of Montreal, Toronto and north-eastern United States), and the lack of local suppliers for specialized equipment and supplies. Most respondents mentioned the importance of their personal contacts within the region, with clients, suppliers, financial and business services. In a few cases, linkages with specific clients outside the region were mentioned. Only in two cases (out of six) were contacts of a scientific nature mentioned explicitly, with researchers at a university and at government laboratories, in and out of the region.

The analysis and comments received during the interviews suggest that the focus of CREDEQ shifted over time from the incubation of technology-based start-ups (technology incubation) to pure business incubation, CREDEQ becoming more a multi-tenant building in the late 1990s than an actual incubator, competing on price rather than on access to researchers, technology and knowledge networks. It lost its competitive advantage with the end of its government subsidies.

Conclusions

Overall, the Quebec Metropolitan area is a really interesting example of a medium-sized government city, relatively isolated geographically, with a good research university and research hospital and R&D activities concentrated in a few niche areas. The region has been able to develop a successful technology base in those niche areas with good early growth. It has not, however, been able to reach the critical size needed to attract large multinational firms in those areas and to develop a business environment conducive to high growth. Four stages can be distinguished in its technological evolution in the past 30 years:

- Stage one (early 1980s): Pro-active decision by local government, the local university, and the business sector to develop an infrastructure supportive of local R&D and technology commercialization activities.
- Stage two (late 1980s and early 1990s): Early growth of research and technology commercialization activities (established firms and spin-offs) in optics, information technologies and health sciences; the domains of expertise in the local university and teaching hospital;
- Stage three (slow-down in the mid-1990s): The region not being able to attract significant external investment and large firms in its niche technology sectors;
- Stage four (late 1990s): Resumption of growth in the high-tech sector in optics, photonics and information technologies, but with a rate of growth significantly lower than in larger centers such as Montreal, Toronto and Ottawa.

What is the potential of the Quebec City metropolitan area as a TIP? It has a number of advantages: a relatively well educated workforce; an excellent university and research hospital, source of world class research and qualified manpower; several prominent research institutes and laboratories; a number of small and medium-sized advanced technology firms in well focused technology niches; the government of the Province of Quebec which spends a significant share of its R&D money in the region; and a well recognized quality of life in a region known for tourism, its cultural activities, and its 'old world' charm, in spite of long and cold winters. The region has, however, not reached the critical mass needed to attract external scientists, entrepreneurs and venture capital. Its culture still seems to be that of a nice and pleasant government town, the entrepreneurial spirit and active networking characteristic of a high growth knowledge region not having reached the level needed for high growth. Except for a change in local and provincial policy, it is therefore unlikely that the region's advanced technology sector will change much in the foreseeable future, continuing its slow growth in a few well-defined niche sectors and trying not to lose advanced technology firms to larger (albeit more expensive) TIPs.

3.2.3 The Saskatoon Metropolitan Area

With 234 000 inhabitants in 2004, Saskatoon is the largest metropolitan area in

Saskatchewan.[58] It is located about 300 kilometers north of the US border, 800 kilometers west of Winnipeg and 500 kilometers east of Edmonton. The other major urban center in that mainly agricultural province is Regina, the provincial capital with close to 200 000 inhabitants. Saskatoon likes to be known as 'Canada's Science City' because of its active biotechnology, food processing, health and advanced technology research activities. In its 2004 Competitive Alternatives Study, the consulting firm KPMG 'ranked Saskatoon second in overall cost-effectiveness among major North American Midwest cities. It has also placed first for R&D in biomedical, clinical trials, and pharmaceutical manufacturing. A biomedical R&D operation in Saskatchewan is 21.7 per cent more cost-effective than in the average US location. Saskatchewan cities are also ranked among the best for location-sensitive costs.'[59]

The Province of Saskatchewan is located in Canada's mid-west region, 'the Prairies'. It had, in 2004, a total population of 996 124 inhabitants and a resource-based economy highly dependent on agriculture (mostly wheat and other grains, livestock, 7.5 per cent of its GDP in 2001), and on oil and mining (potash, uranium, crude oil, 17.4 per cent of its GDP), with limited manufacturing activities (only 7.5 per cent of the provincial GDP).[60] Between 1993 and 2003, the provincial real GDP grew at an average annual rate of 2.2 per cent boosted by favorable conditions for oil and potash exports (but still below the 3.5 per cent average growth rate for Canada as a whole). In 2003, Saskatchewan's GDP per capita was 4 per cent below the national average, at C$36 723 compared with C$38 495 for Canada.[61] For many years, the province has experienced the lowest population growth rate among all Canadian provinces except for Newfoundland, many of its young people emigrating to Alberta, Ontario or British Columbia.

Regional technology infrastructure

With a R&D/GDP ratio of 1.2 in 2002[62] (C$421 per capita), Saskatchewan ranked 5th among provinces in Canada in research intensity, up from 7th in 1995. A major portion of the R&D is performed in universities and research hospitals (61.8 per cent in 2002 compared with 33.4 per cent for all of Canada),[63] 23.2 per cent being carried out by industry (55.4 per cent for Canada), 12.6 per cent by the federal government and 2.4 per cent by the provincial government (9.8 per cent and 1.4 per cent respectively for all of Canada). Public sector research activities tend to reflect the province's economic profile with emphasis on agriculture and biotechnology – federally through the Agriculture and Agri-Food Canada's research center, and the NRC's Plant Biotechnology Institute in Saskatoon; and provincially through the activities of the Saskatchewan Research Council. About 90 per cent of the research activity of the province takes place in Saskatoon. A major per centage of the province's manufacturing activity is also located there.

Scientific activity in Saskatoon started by chance. In the early twentieth century, the provincial government gave St. Albert, a large agricultural center and Saskatchewan's largest town at the time, the choice between a large correctional center and a university. St. Albert chose the correctional center for its short-term pay-off and the university went to Saskatoon where it was officially established in 1907. Over time, the University of Saskatchewan became Saskatoon's main economic engine, the local institution which most impacted the city's economic development. It is still the city's largest employer. The economic impact of the University of Saskatchewan can be attributed in large part to the excellence of its research activities. Agricultural research activities accelerated during the Second World War under a government mandate to develop a variety of the oilseed rape plant to produce industrial grade oil. This was successful and resulted in significant rapeseed oil production in the Canadian Prairies. Further research led to the development of Canola (= 'Canadian + oil'), a rape plant developed through careful plant breeding and producing an edible oil of very high quality. Canola's success as a new food crop firmly established the reputation of Saskatoon as a world class research center in new plant technologies, ag-biotechnology, and now nutraceuticals (plant-based pharmaceuticals) and vaccines. Excellent research in engineering at the University of Saskatchewan, especially space sciences research in the 1940s and 1950s and research done for local mining and manufacturing firms, also led to the development of leading companies in telecommunications and systems.

The University of Saskatchewan and the Plant Biotech Institute (A National Research Council research institute) located on its campus have been key factors in the development of a technology cluster in Saskatoon, in the 1980s and 1990s. Such development resulted from the provincial government decision in the mid-1970s to create a science park, Innovation Place, which soon became the home of a number of federal and provincial research laboratories and institutes. Thanks to the strong pull of Innovation Place and of its solid research base, advanced technology industries began to develop in Saskatoon in the early 1980s. A healthy agricultural biotechnology industry has developed (plant and animal breeding, pesticides, plant biotechnology with particular emphasis on canola-related activities) which grew from one firm in 1992 to over 30 firms in Saskatoon 'including BASF Canada, Bayer CropScience Canada, Dow AgroSciences, and Monsanto (international companies), as well as Philom Bios Inc. and Bioriginal Food & Science Corp. (local start-up companies)',[64] which make up about 30 per cent of the Canadian 'ag-biotech' industry. And, from about 30 firms in 1981, the Information Technologies sector now employs 8000 people in nearly 200 firms with C$1.6 billion of sales annually[65] with special expertise in telecommunications, satellite control, instrumentation, robotics and geographic information systems.[66] The opening in October 2004 of 'the Canadian Light Source' synchrotron on the University of Saskatchewan campus further adds to its solid research infrastructure.

In 1998, the University of Saskatchewan had 13 200 full-time undergraduate students and 1520 full-time graduate students, as well as 4000 part-time students (3500 undergraduate, 500 graduate) enroled in its 13 colleges and schools (agriculture, arts and science, commerce, dentistry, education, engineering, graduate studies, kinesiology, law, medicine, nursing, pharmacy, veterinary medicine, and extension division). With its six life science faculties, a major research hospital and a number of key institutes in agricultural biotechnology, its potential in ag-biotech is very high. Its research budget of C$116 million (2002–2003) represents a major share of the estimated C$419 million spent on R&D activities in all of Saskatchewan. In a recent survey (Harms et al., 2001), University of Saskatchewan Technologies Inc (UST), the wholly owned company founded in 1991 to commercialize University of Saskatchewan technologies, identified 33 spin-off firms: seven early stage, 17 active, two merged/acquired and seven inactive, in life sciences (18 firms), physical sciences (10 firms), information technologies (five firms), with a total of 1383 employees and C$190 million in combined sales. Twelve of those firms are located in the park, including SED Systems with its 250 employees, University of Saskatoon's first and largest spin-off, created in 1972. The fact that those spin-off firms did not seem to have made much use of the services of UST led Margaret Barker to conclude that 'our results do not point to the University of Saskatchewan as being the leader in regional technology-based economic development, rather, it appears to be an important catalyst in that development' (Barker, 1996, section 3.1.5).[67]

Science park and incubators

As noted previously, the provincial government decided in the mid-1970s to create a science park, Innovation Place, as an economic development tool to support the development of a technology-based community in association with the university, encourage the commercialization of university knowledge, and generally provide a good home for scientific activities and their commercialization. Innovation Place was created in 1977 by the Provincial Government on 120 acres of land owned by the university and leased to the government, not far from NRC's Plant Biotechnology Institute. Responsibility for the development was given to the Saskatchewan Economic Development Corporation, a provincial crown corporation (which became the Saskatchewan Opportunities Corporation in 1995 and was wound down in 2002). The University of Saskatchewan has three representatives on the park's board of directors (with veto rights for tenant approval), the six other members being named by Innovation Place. Actual development of the park started in 1980 with the construction of two buildings, one by the Saskatchewan Research Council, the other, a multi-tenant building, by Innovation Place. Its first five tenants were provincial entities. In the mid-1980s, a research laboratory was built by Environment Canada and, in 1989, Innovation Place built a second multi-tenant building. By 1990, Innovation Place was considered to be one of the 20 most successful university-related research parks in the world. Most (federal and provincial) research institutes in the region are located in the park or adjacent to it, as well as many agricultural biotechnology firms and a large number of support service companies. The park currently hosts 120 organizations, employing 2000 persons in 18 buildings. It contributes over C$248 million annually to the local economy.[68] Part of its success as a growth pole for this isolated agricultural region comes from its well defined focus on agricultural biotechnology and related areas (in 1997, close to 40 per cent of its activity was in agricultural research, 25 per cent in information technologies, 17 per cent related to the environment, 5 per cent in medical/pharmaceutical, the rest being divided among resource research, environment and other sectors),[69] from the continuing solid support of the provincial government since inception, the colocation of provincial and federal research institutions in agricultural and biotechnology research, and the research activities of the University of Saskatchewan. Specific key success factors included the presence, since the 1950s, of the Plant Biotech Institute (formerly known as the NRC Western Canada Laboratories, located on the campus of the university) with extensive research greenhouses built at Innovation Place. Another influence was the creation in 1989 of Ag-West Biotech Inc, a provincial not-for-profit organization dedicated to the development of 'strong, vibrant, and profitable agbiotech industries',[70] facilitating linkages between research institutions (provincial and federal laboratories, and universities) with industry and within industry, helping with the commercialization of biotech. The opening in 1992 by TRLabs of a laboratory fostering research and commercialization partnerships with industry was another key success factor for Innovation Place.

There was no formal brick and mortar technology incubator in Saskatoon or at Innovation Place until the opening of an industry partnership facility at NRC's Plant Biotechnology Institute in 2004 and venture capital resources in town are still limited. However, Innovation Place provides an environment supportive of start-ups and spin-offs in spite of its relatively high rental cost. Its mix of large and small firms, its concentration of research and business activities, the presence in the park of all the business, financial, legal and other services needed by small businesses, the high quality of the facilities, the ease of networking and social gathering in some of the multi-tenant buildings, and the flexibility shown by Innovation Place management in some cases of delayed rent payments, all make it a supportive place for start-ups without creating the dependency that sometimes results from subsidized rents or the trauma of having to move at the end of a fixed period in incubation.

Characteristics of TBFs

To learn more about the innovation activities of local technology-based firms (TBFs) and the impact of the park/region on their activities, questionnaires were distributed to all Innovation Place tenants in the Spring of 1999. In spite of the support of Innovation Place's management who actually did the mailing, only seven responses were received, from two associations/consortia, two research laboratories, and three product-oriented technology firms (a distribution fairly representative of the park tenant population). In the summer of 2002, a random sample of 20 other tenants led to telephone interviews or faxed responses from six firms (no feedback was received from four of the 10 product/service firms in the sample; other organizations in the sample included six foundations, government offices or administrative offices, three consulting firms, one retail outlet and one financial organization). The nine TBFs which responded to the survey (1999 and 2002) represent about 15 per cent of the product-oriented firms in the park and are therefore not very representative.

The nine respondents are fairly typical small technology-based firms in bio-industries (five), IT, software and related activities (four), taking advantage of the best rental business space in town, with easy access to research laboratories and scientists. They are relatively young (only two firms are over eight years old), with three to 100 employees, about half with no sales yet (several firms in pre-clinical or clinical tests), the others with up to a few million dollars in sales, mostly active in export markets. All respondents reported one to three product innovations, a few process innovations and an occasional managerial innovation. Innovations were mostly market-driven (70 per cent of the time on average) and mainly developed in-house (90 per cent to 100 per cent in-house for the bio-industry firms, 50 per cent to 90 per cent in-house for the IT/software firms, typical behavior for technology start-ups). Patents were used by all bio-industry firms to protect their intellectual property (five to over 20 patents, generally their own, in one case licensed from the university); only one IT firm had patents.

All respondents valued Innovation Place as a very supportive incubation space. Several commented that there was no need for a formal brick-and-mortar incubator in the park, the park and its tenants offering all the services and networking opportunities they needed. Among the advantages most often cited, we can note:

- the quality of the infrastructure (buildings, security, laboratories, telecommunications);

- the proximity and easy access to R&D labs, the university, research hospitals;
- opportunities for business and social networking with others in the same field or facing similar business challenges;
- on-site availability of business and technical services;
- a recognized prestigious address;
- the cost of Innovation Place was a 'negative' factor named by three respondents, the lack of critical mass in his field (at Innovation Place and in the region) being another negative factor given by one respondent.

As a region, the Saskatoon metropolitan area was judged very positively for its research intensity, excellent university, excellent quality of life, availability of well trained personnel, and low cost. Lack of venture capital, distance to market, and poor airline connections were, however, negative factors noted by almost all the respondents who felt that they may have to move to grow.

Conclusion

Overall, the Saskatoon Metropolitan area is another interesting case, that of a relatively small geographically isolated city with a good research university which launched successfully in the late 1970s an aggressive niche strategy to develop a technologically-intensive community. Success came from the very solid cooperation between the federal and provincial governments and the university which led to the development of a very solidly focused TIP. The role of Innovation Place as a local catalyst for research and commercialization activities must also be noted.

The development of the region as a world-renowned ag-biotech center can be divided into five stages:

- Stage one (mid-1970s): Decision by the provincial government to support the development of a technologically-based community associated with the university, by creating a science park to provide a good home for scientific activities in ag-biotech and related fields and their commercialization;
- Stage two (in the 1980s): Development of a solid research infrastructure with the building of a number of provincial and federal research laboratories adding to the ag-biotech base already in place.
- Stage three (events of 1989, 1992): Rapid development of the local industrial base with two events, two milestones which contributed to local growth: the creation by the provincial government in 1989 of AgBiotech, a not-for-profit organization dedicated to the development of a vibrant ag-biotech industry; and, the installation at Innovation Place

in 1992 of TRLabs, a leading not-for-profit information and telecommunication technology research consortium attracted by the region's solid research infrastructure.

- Stage four (in the 1990s): Successful expansion of the local industrial-research base in ag-biotech and telecommunications/software, riding the mid- to late-1990s technology cycle, and an opportunity for a region that was ready at the right time;
- Stage five (early 2000s): Reduced growth following the end of telecommunication and dot.com cycles and also resulting from mixed market feelings for genetically modified products. Lack of venture capital and distance to larger technology centers are additional constraints on Saskatoon's continuous growth.

Saskatoon has many of the elements needed for sustainability: a good quality workforce and a very good research university, a solid research base in its two leading sectors, ag-biotech and IT, a number of private firms active in those two sectors (branches of multinationals and locally grown firms), a good local culture supportive of research-based entrepreneurial activities, a very committed regional development organization, and a good quality of life. On the negative sides are the relative small size of the city, its geographic isolation and its hard winters, and the lack of venture capital. Because of the quality of the research being carried out locally and its intensity, it is expected that Saskatoon will continue to be a world-class ag-biotech research center. Further developing and keeping its advanced technology firms will, however, be a challenge as the city competes with larger centers that have larger clusters of firms in similar domains and with more venture capital.

3.2.4 The Calgary Metropolitan Area

Located in the Province of Alberta, at the foot of the Rocky Mountains in western Canada, Calgary is the center of Canada's oil industry. It is a city of one million highly skilled inhabitants (2002), with the second highest number of large firms' head offices in Canada (after Toronto).[71] In the 1970s, its economy was almost exclusively driven by oil-related activities which included the radio communication systems needed to communicate with far-removed oil fields and the advanced software and imaging capabilities needed for geological surveys for oil exploration. It is one of the fastest growing cities in Canada with good cultural activities, a high quality of life, strong entrepreneurial spirit, and relatively moderate cost of living (compared with Toronto and Vancouver). Thanks to its oil revenue, Alberta has the lowest corporate and personal tax rates among all Canadian provinces. Calgary has a well educated labor force (18.53 per cent of population aged 15 and older with a university degree, compared with 22.6 per cent in Ottawa, the highest in Canada, and Toronto, the second highest at 18.9 per cent), and a wealthy population (average annual household income of C$66 900 and 8.10 per cent of households 'very wealthy', compared with C$68 400 and 14.8 per cent respectively for Toronto, Canada's wealthiest metropolitan area.[72]

Regional technology infrastructure

In the late 1970s and early 1980s, as oil and gas firms were going through a recession, the provincial government decided that it was important to diversify the provincial economy. This led to the creation of the Calgary Research and Development Authority (CR&DA now Calgary Technologies Inc. (CTI)),[73] a not-for-profit tripartite initiative of the University of Calgary, the local Chamber of Commerce and the City of Calgary created in 1981. This also led in 1982 to the creation of NovAtel Communications, a joint venture of Nova Corporation, a local natural gas company with a very large radiotelephone system, and of the Alberta Government Telephone company (now Telus). NovAtel became a very large Calgary-based manufacturer of cellular telephones, the largest in the world initially before being overtaken by Motorola, and also developed expertise in global positioning systems. During the 1990s, NovAtel expanded its manufacturing and systems capacity. By 1990, it had 1900 employees, 858 located in Calgary (Langford et al., 2002, p. 3), making it the largest high-tech employer in town. At that time, other large technology firms in town included Nortel (telephone equipment manufacturing) and Computing Devices (contract electronics). There were also a number of smaller technology companies.

To make the region more attractive to high technology industries, CR&DA started in 1983 to develop the University of Calgary Research Park on land owned by the government leased to the city of Calgary. The park (currently with seven buildings, 1.2 million square feet of space and over 3000 jobs)[74] has no sectoral focus. It now houses several private and government research laboratories, as well as Discovery Place, a large university-owned multi-tenant facility. CR&DA's Technology Enterprise Centre, created in 1985 to provide assistance to technology-intensive firms, also operates two very successful business incubators, one located in the park and the other in the same building as the Alberta Research Council in the center of Calgary. CR&DA's objective, in partnership with the University of Calgary, was to make the region attractive to technology investments. In 1992, CR&DA launched 'Infoport', a project designed to encourage networking, attract venture capital and develop external markets.

In 1989, the Government of Alberta encouraged the creation in Calgary by TRLabs of a wireless laboratory which would work with the University of Calgary (Langford, 2002, p. 7), thus enhancing Calgary's position as a center for wireless research activities. TRLabs is an Alberta-based not-for-profit university–industry–government telecommunication research consortium created in 1986 by the government of Alberta, Nortel and the University of Alberta.

In the late 1980s, NovAtel went through serious business difficulties.[75] In 1992, Nortel agreed to purchase its systems business to create a large wireless center of excellence in Calgary, the beginning of Calgary's wireless cluster, putting the city in the limelight as a recognized high-tech region. Changes at NovAtel released a large number of highly skilled engineers and scientists, many potential technology entrepreneurs and also a bonanza for other technology firms, an important boost for the nascent wireless and telecommunications cluster. In 2000, the wireless and telecommunications cluster had over 100 firms, over 12 000 jobs, and had attracted a number of multinationals (Master and Reichert, 2000). It is estimated that more than half of those firms had been created since 1995 (Langford, 2002, p. 8). Other technology sectors of importance included IT software (800 firms, 10 000 jobs), IT services (300 firms, 12 000 jobs), geomatics, of special importance to oil firms (80 firms, 3500 employees), multimedia (85 firms, 2500 jobs), and electronics manufacturing (50 firms, 7000 jobs). Altogether, it is estimated that in 2000, Calgary had over 55 000 high-tech jobs, 1200 high-tech firms with over C$10 billion in sales (Master and Reichert, 2000) and that, in 2000, about C$200 million of venture capital was invested in local firms.[76]

Science park and incubators

As noted above, CR&DA started in 1983 to market the 125-acre University of Calgary Research Park located close to the university's faculties of science and of engineering. Formal technology incubation activities in Calgary stated in 1985 with the creation by CR&DA of the Technology Enterprise Centre (TEC) to operate two small incubators,[77] one providing support and guidance to firms renting space in Discovery Place (a 65 000 square foot multi-tenant building located in the park, expanded to 105 000 square feet in 2000), and the other one, now closed, which used to be colocated with the Alberta research council in the 'Digital Building', in downtown Calgary. By the end of 1997, it was estimated that about 280 companies had participated in TEC activities and over 700 jobs had been created (Croft, 1997), mostly one- and two-person operations. The number of companies having benefited from TEC support had grown to an estimated 350 by 2000, many firms having three to four employees.[78] The University of Calgary is not directly involved with those incubators, nor is the organization which manages the intellectual property of the University of Calgary, University Technology International. Indirectly, however, the Faculty of Management of the University of Calgary has had a very positive impact on local entrepreneurial spirit and on new technology venture creation through its New Venture Group, a leader in entrepreneurship research for over two decades, and its innovative entrepreneurship MBA program (Chrisman, 1994).[79] In 2000, a new incubating organization, Innocentre, a very successful Montreal-based not-for-profit virtual incubator assisting high potential start-ups, opened an office in Calgary to provide support to promising technology start-ups. There are also in town several small venture capital organizations which incubate the ventures they finance.

As noted above, CTI operates Discovery Place, a state-of-the-art 105 000 square feet multi-tenant building located in the University of Calgary Research Park. Discovery Place offers flexible space in a beautifully designed building with a superior communication infrastructure, common services and meeting rooms, access to business counseling, assistance in accessing technology, financing, clients, suppliers, and a sense of community to facilitate networking and sharing. Its objective is to provide an environment conducive to start-ups and other tenants, and to encourage growth. In its early years, CTI (formerly CR&DA) and Discovery Place were financially supported by various levels of government and the city of Calgary. In the late 1990s, Discovery Place had an operating surplus. Its expansion in 2000 from 65 000 square feet to 105 000 square feet was financed by the province (about C$1.5 million), private donors (about C$1.5 million), the local economic development authorities (C$0.5 million) and debt (C$4 million).[80] The debt will be financed from rental income and sale of services. Rents are market driven and on the high side because of the quality of the facilities and the prime location. In 2000, Discovery Place had no formal entry criteria except that candidates were expected to be technology based (generally at or close to the commercialization stage; no manufacturing allowed on-site) or to be service providers to tech-based firms (intellectual property law, business services) and to add value to the tenants already in the building. There were no formal exit rules for tenants or a limit on the duration of their stay at Discovery Place, with some leases for up to ten years. In 2000, CTI was working with 75 to 80 organizations, about 50 located at Discovery Place and the others as members of the 'corporate identities' program being able to use Discovery Place's address, and having access to its common facilities and services for a monthly fee. About 60 per cent of the firms involved with CTI, at Discovery Place and outside, are knowledge-producing organizations, the balance being private or public service organizations.[81] Start-up entrepreneurs at Discovery Place come from a variety of places or previous employments. Few actually come from the University of Calgary even if some university professors are involved privately as consultants and some students work as employees or on academic student projects sponsored by tenant firms (class projects, MBA graduation projects). Discovery Place organizes occasional seminars and networking events, offers business counseling and facilitates access to clients, suppliers, venture capital and other sources of financing to those who request it. However, most of its tenants seem to be there more for its well designed congenial rental space and superior telecommunication infrastructure than for the value-added services typically offered by formal technology incubators.

Characteristics of TBFs

Information on technology incubation in Calgary was collected in stages:

- In 1996–97 for a previous research project on technology incubation (Doutriaux, 1999): a short two-page questionnaire (15 minutes average completion time) focusing on start-up characteristics and the technology used, reasons for joining the incubator and its advantages and disadvantages, linkages with the university had then been sent in 1996 to the tenants of the Technology Enterprise Centre. Responses were returned using a toll-free fax number. Eleven questionnaires were returned (a 33 per cent response rate), nine from technology-based firms, two from professional service organizations.
- In addition, for this project, short visits were made to Calgary in January 1999 and again in August 2000, to interview professionals involved with technology transfer and technology incubation in Calgary (at Calgary Technologies Inc (CTI), the University of Calgary, and other organizations). Those interviews provided the basic information for this section. A survey was also organized using the Monarca questionnaire to learn more about the innovativeness and the incubation experience of local TBFs. In spite of the support of CTI which helped with the distribution of the Monarca questionnaire to its incubatees (product companies only; service organizations and testing laboratories were not targeted), only two interviews/questionnaires were completed in 1999 with the original questionnaire and one more was received in 2000, a shorter four-page version, after a telephone and e-mail invitation to participate sent to a small number of incubatees. This very poor response rate can be blamed on survey fatigue, TBFs having been the focus of too many studies in the late 1990s, and also on the length of questions of the Monarca questionnaire.

The analysis that follows is based on the twelve questionnaires coming from technology-based firms, the profiles of the 1996–1997 and of the 1999–2000 respondents and their comments being very similar.

Six respondents were software developers, four were in instrumentation, electronic systems, and information technology, and two were in environmental and renewable energy activities. Four firms were founded in the 1990s, seven in the 1980s and one many years before. Median age at time of entry at

Discovery Place was five years (0 to 11 years, with one older firm) and at interview time, they had been between less than a year and over ten years at Discovery Place. Sales at entry were between zero and C$300 000; current sales between C$150 000 and C$3 million, half of the firm reporting high growth (above 30 per cent/year) since entry, a growth rate not uncommon in the late 1990s. The firms interviewed in 1999-2000 reported one to two product innovation(s) and 1 to 2 service innovation(s), their innovations being driven slightly more by the market (30 per cent to 80 per cent) than by technology (except for one case of pure technology product innovation) and having been developed mostly but not exclusively in-house (30 per cent to 100 per cent). Half of the 1996–97 respondents had said they went through hard times before entering Discovery Place; all respondents said that their competitive situation and sales had improved since their entry.

For a majority of the twelve respondents to the two surveys, the main advantage of Discovery Place is that it is a well designed multi-tenant building in a good location, with very good security, an excellent telecommunication infrastructure, and access to shared office services and common spaces. Networking opportunities with other firms were noted as an advantage by about half of the firms. About a quarter of the respondents mentioned proximity/contacts with the university, its R&D activities, its professors, its students (as a well qualified, well trained, reasonably priced source of labor) as an advantage of Discovery Place. Image and a prestigious address were also mentioned by a quarter of the respondents. It is worth noting that no firm selected 'incubation' on the questionnaire as a reason for its entry into Discovery Place, a result perfectly consistent with the views expressed by several professionals interviewed in Calgary that Discovery Place is primarily an excellent multi-tenant building. The 1999–2000 respondents noted the region's excellent labor market, telecommunication infrastructure, quality of life, but complained about the distance to major US markets.

Conclusion

The Calgary Metropolitan area offers an interesting example of the successful diversification of a large city from a wealthy oil-producing base and related R&D activities into an advanced technology region under the leadership of the provincial government, and in partnership with the local research university and the private sector.

Five stages can be observed in Calgary's high-tech development:

- Stage one (late 1970s to early 1980s): decision by the government of Alberta to take measures to build a diversified technology industry taking advantage of local expertise in telecommunications and geomatics and of the city's high quality of life; creation of CR&DA, a local development authority, in partnership with the University of Calgary.

- Stage two (1980s): development of telecommunication and wireless research and manufacturing activities with the creation and growth of NovAtel; significant expansion of the pool of expertise in that domain, with a strong regional inflow of engineers and scientists, and expanded activities at the University of Calgary.
- Stage three (1989–1992): transition and restructuring with the sale of NovAtel's system business to Nortel, and with Nortel's significant expansion of its wireless activities in town. Creation by CR&DA of Infoport to boost the region's image in high-tech and to build on the visibility brought to the region by the 1988 winter Olympic games.
- Stage four (1990s): growth of wireless, geomatics, software and other high-tech activities, riding the tech wave, with the expansion of the activities of large firms in town (Nortel, Telus, IBM) and the growth of successful start-ups. Development of specialized business, financial, and technical services in parallel with the high-tech development.
- Stage five (early 2000s): consolidation, with the slowdown of the telecommunication and related industries, access to venture capital and to the market being seen by many local professionals as a problem which could limit the region's growth.

Overall, Calgary has most of the elements needed for sustainability in advanced technology: a well trained workforce and well trained researchers, a good research university with good linkages to the business world, a number of large TBFs (branches of multinationals and locally grown firms), a dynamic entrepreneurial culture with public and private entities actively promoting the region and encouraging connectivity and networking. High quality of life and low taxes (relative to the rest of Canada) are other positive factors. Current challenges to sustain growth include lack of venture capital, weaknesses in technology commercialization (technology sales skills and distance to markets), and the retention of growing firms. Indeed, as has been observed in several other Canadian TIPs, in times of sectoral slowdown as experienced in the early 2000s in telecommunications and photonics, successful start-ups which have attracted significant foreign capital are easily moved out of the country when their foreign owners decide to consolidate their operations in a reduced number of geographical locations.

3.2.5 Conclusions

In three regions (Quebec, Saskatoon and Calgary), the high-tech development process can be linked to a clear desire by governments (local, provincial, with federal support) in partnership with local universities to diversify the local economy (building directly on the areas of expertise of their universities in Saskatoon and Quebec, and on the existing advanced technology base in Calgary). This process led to the creation of science parks to provide visibility to their initiatives, attract public and private research organizations and high-tech firms, and facilitate the creation and development of home-grown TBFs. In the fourth region (Ottawa), high-tech development was more opportunistic and started with private sector initiatives, the region acting as a dynamic incubating milieu. In that region, it is only after provincial and local authorities, in partnership with the University of Ottawa and its teaching hospitals, decided to develop a new advanced technology cluster in life sciences that a 'life science technology park' was created, the first science park in the region. The science parks in all four regions are managed by independent non-profit or government organizations and have representatives of local universities, three levels of governments and the private sector on their boards.

Key to the success of the development of all these regions have been the following:

1. Their solid research base: very high research intensity in many domains in the Ottawa region with particular focus on telecommunications (CRC (federal lab.) and Nortel (private lab.)), microelectronics, computers and life sciences (the two local universities) and many other domains (NRC, federal lab.); high research intensity in niche domains in the other regions (Saskatoon in plant biology and ag-biotech in particular at the university and in a number of federal and provincial research institutes; Quebec in optics, information technologies, forestry and life sciences in particular at the university and in several public research laboratories; Calgary in engineering, wireless and information technologies at the university and private firms.
2. Federal and provincial government programs, and local networking activities to encourage research partnerships between industry and public and university laboratories, and to facilitate technology transfer with special focus on small and medium sized firms.
3. The presence of a proactive development organization (a 'champion') to facilitate linkages, and encourage cooperation between governments, universities and private firms: public organizations (Ag-Biotech in Saskatoon) or consortia of public, university and private interests (CR&DA in Calgary, OCRI in Ottawa). In Quebec, late 1990s high-tech development seems to have suffered at least in part from a lack of consensus or effective cooperation between all concerned groups.

4. The presence of one or more large research/manufacturing private firms, a large 'anchor organization', with a large number of engineers and scientists, providing visibility to the region in its field, a source of potential spin-offs, a potential client for start-ups, a role model. In Ottawa, Nortel has been that key anchor firm, followed by Mitel and Newbridge. In Calgary, NovAtel, followed by Nortel have played that role. In Saskatoon and Quebec, a lower rate of new technology firm creation may be explained in part by the absence of a very large private sector 'anchor firm', that role having been played by one or two successful local start-ups.
5. The presence in each region of at least one good research university as source of highly trained manpower (required) and good research (useful but not as important, if there are other proactive public and private research laboratories in the region). Post-secondary technical colleges are also very important as a source of highly trained manpower.
6. Good quality of life and good physical communications with the United States and the rest of the world (lack of easy physical communications with the USA has been a limiting factor for Saskatoon and Quebec).

In most regions, business services (management consulting, technology assessment, technology marketing, legal firms dealing with corporate and intellectual property issues), technical services (such as precision machining, prototyping, precision molding and testing), and the supply of financial resources (angel investors, venture capital, banks and other sources of capital) have evolved in parallel with local high-tech development, although generally lagging behind that development and therefore slowing it down.

In all four cases, key to sustainability and continuous advanced technology growth is clearly the continuation of a high level of research activities and of good collaboration between the three levels of government, universities and the private sector, as well as measures designed to keep the regions competitive in terms of quality of life and physical communications (airport and air links in particular). The continuous nurturing by regional stakeholders of a local entrepreneurial culture supportive of connectivity and networking and leading to the creation of a large number of technology-based start-ups is also very important (Brouard et al., 2004). One major challenge for the four Canadian TIPs surveyed is to keep their home-grown start-ups and to create conditions that will see them growing locally. It is therefore important to develop public policy measures designed to increase the supply of local venture capital, to develop, attract and retain top managerial talent (technology leaders with a vision in particular), and to develop, attract and retain the top expertise in technology sales and marketing that will help local companies be more competitive in the commercialization process.

NOTES

1. Statistics Canada, CANSIN Matrices 6367-6368 and 6408-6409.
2. According to the 1996 Census of the Canadian population, in 1996, 20 per cent of the 25–29-year-olds in Canada had a university-level education, 32 per cent had a college/trade-level education, 29 per cent had completed high school, and 18 per cent had less than a high school education (1999).
3. Estimated from 1996 Census data, http://www.statcan.ca/english/census96/apr14/educ.htm.
4. Based on Statistics Canada data, http://www.statcan.ca/english/Pgdb/econ05.htm.
5. All figures in the chapter are in Canadian dollars (C$) unless otherwise indicated (in late 2004, one Canadian dollar was equal to about 0.82 US dollar).
6. The fact that, in 2002, the Province of Ontario produced more cars than the State of Michigan is little known.
7. The trade deficit in machinery and equipment would be even higher without the country's C$5 billion trade surplus in aircrafts and related equipment.
8. Canadian Statistics on International Trade, http://www.statcan.ca/english/Pgdb/gblec02a.htm.
9. Telecommunication equipment, electronics parts and components (semiconductors), computer equipment, communication wires and cables, instrumentation, consumer electronics, communication services, software and consumer services, ICT wholesaling, office machinery rentals and leasing.
10. 'The ICT Sector in Canada', August 2001, Industry Canada, http://strategis.ic.gc.ca/pics/it/sp1199e.pdf.
11. ICT Sector Intramural R&D Expenditures, 2004 Intentions, Industry Canada, Information and Communications Technologies, http://strategis.ic.gc.ca/epic/internet/inict-tic.nsf/en/h_it05385e.html, read on 25 February, 2005.
12. 'The ICT Sector in Canada', op.cit., p. 3.
13. Overview of Biotechnology, http://strategis.ic.gc.ca/SSG/bo01544e.html. 8 August, 2001.
14. Statistics Canada, 2004, Service Bulletin Science Statistics, Cat 88-001-X1E, vol 28, n.9, August 2004, Table 2.
15. Government-owned corporations.
16. Canadian Institute for Health Research since June 2000 (http://www.cihr.ca/).
17. For comparative purposes, other selected GERD/GDP 2001 ratios: Sweden (4.27 per cent, Japan (3.07 per cent), Finland (3.41 per cent), Germany (2.51 per cent), France (2.23 per cent), the UK (1.86 per cent), Italy (1.11 per cent) (OECD 2004c, Table 2).
18. All values in current US Purchasing Power Parities adjusted dollars, values which reflect local purchasing powers: <http://www.oecd.org/department/0,2688,en_2649_34357_1_1_1_1_1,00.html>.
19. C$4492 million for expenditures on R&D alone. In 2002–03; the Federal Government total expenditures on Science and Technology (science and R&D) were C$8014 million (Statistics Canada, 88001-XIE, vol. 28, n.11, November 2004, table 2).
20. Statistics Canada, 88-001-X1E, 28(9), August 2004, Table 1.
21. In comparison, Japan had 9.9 researchers per thousand persons in its labor force, with 6.8 in France, 6.7 in Germany, and 5.5 (1998) in the UK (OECD, 2004c, Table 8).
22. Estimated from number of S&E publications from National Science Foundation NSF (2004) Appendix table 5-35, and number of researchers from OECD (2004c) Table 7.
23. Patents registered with the European Patent Office, the US Patent and Trademark Office and the Japanese Patent Office, a measure developed by the OECD to facilitate international comparisons.
24. OECD 2004b, p. 38.
25. OECD 2004b, p. 47.
26. Ibid.
27. OECD, 1997c, p. 9.
28. Statistics Canada, 2004, Service Bulletin Science Statistics, Cat 88-001-X1E, vol 28, n.9, August 2004.

29. http://www.aucc.ca/en/aboutindex.html.
30. http://www.accc.ca/english/Colleges/membership_list.cfm.
31. http://www.nce.gc.ca/en/netseng.htm.
32. Statistics Canada, 2004, Service Bulletin Science Statistics, Cat 88-001-X1E, vol 28, n.5, May 2004, Provincial research organizations.
33. www.nrc-cnrc.gc.ca.
34. http://precarn.ca/Corporate/corporate.cfm; http://www.canarie.ca/about/about.html.
35. Networks of researchers focusing on the same field or the same health issue.
36. http://irap-pari.nrc-cnrc.gc.ca.
37. From the data base of OCRI, the Ottawa Centre for Research and Innovation, Q1-2003.
38. Now called Ottawa Centre for Research and Innovation, ww.ocri.ca.
39. http://www.researchinfosource.com/top50.html for FY 2000 estimates. See also Doutriaux (2000) for the 1998 percentage estimates.
40. www.ocri.ca.
41. Ottawa Economic Development Corporation (now part of OCRI, www.ocri.ca), '1998 Ottawa facts'.
42. C$29 million for the Quebec portion and C$776 million for the Ontario portion of the Ottawa-Hull metropolitan area, for government R&D expenditures (Stats Canada, cat 88-001-XIB, v.24, n.6, December 2000); Total Science and Technology federal expenditures in the region amount to C$1789 million (Stats Can, cat 88-001-XIB, 25(1)).
43. In 2002, local federal R&D expenditures had grown to C$1015 million (Stats Can, cat 88001-XIE, 28(12), Table 8). In 2003, university R&D stood at C$186 million at the University of Ottawa and C$52 million at Carleton University (Table 3.5).
44. http://www.entrepreneurship.com.
45. The *Ottawa Life* Sciences Technology Park was created in 1992 as a University of Ottawa initiative backed by the provincial government (provided the land) and the city of Ottawa. It failed because of its dependence on the financial support of the university and the local hospitals. It is now owned by the Ontario Development Corporation (ODC), a government development organization, and managed jointly by the ODC, the Life Sciences Council (a public/private partnership representing the health sector's interests and including the University of Ottawa, local hospitals, National Research Council (federal) and a number of private firms), as well as the city of Ottawa.
46. There were a number of very successful start-ups which were spun off, and also several failures. This program was terminated in 1999, Nortel having decided that it was easier to buy successful start-ups with the right technology than to grow its own start-ups to develop the technology.
47. Some of the most successful affiliate firms include Skystone, sold to Cisco in 1997 for US$100 million in cash; Cambrian Systems was sold to Nortel for US$300 million; Extreme Packet Devices sold to PMC-Sierra for US$415 million.
48. About 4000 persons working locally in 100 life science companies and life science. 7000 researchers and scientists in local universities and research centers (Ottawa 2003).
49. OCRI Ottawa Fact Sheet, December 2004, http://www.ocri.ca/ocrimodel/publications/1204_ottawafacts.pdf.
50. The Canadian Encyclopedia, second edition, 1988, 'Quebec City', p. 1805.
51. Gouvernement du Québec, Ministère des régions, région 03 (Capitale Nationale), http://www.infostat.gouv.qc.ca.
52. Table de bord des systèmes d'innovation du Québec, www.mderr.gouv.qc.ca 2004, pages 26, 26, 34, and 36.
53. www.speqm.qc.ca/speqm-fra/cont-quebec-technoregion.html.
54. Statistics Canada, cited by Annie Laurin, in *Le Soleil*, 7 October 2001, 'Pleins feux sur Québec, une capitale à vocation technologique', p. 2.
55. From an interview with a person familiar with CREDEQ, in June 2000.
56. 'Young entrepreneurs create new culture', The Globe and Mail, 15 November, 2000, p. Q12.
57. Répertoire des entreprises ayant bénéficié des services du CREDEQ, January 2000. About half of the questionnaires were mailed directly by the researcher. CREDEQ offered to send the questionnaires directly to the other firms.

58. http://www.city.saskatoon.sk.ca, quick facts, January 2005.
59. http://www.areadevelopment.com/Pages/Features/Feature11.html, 22 January, 2005.
60. Government of Saskatchewan Bureau of Statistics, www.goc.sask.ca/bureau.stats, January 2005.
61. Ibid.
62. Statistics Canada Catalog 88-001 XIE, 28(12), table 4, December 2004.
63. Ibid, Table 8.
64. Saskatoon Regional Economic Development Authority, www.sreda.com/science-city_biotechnology.php, January 2005.
65. http://www.gov.sk.ca/econdev/investment/sixsctrs/technology/sector.shtm.
66. Part of this growth is due to the University of Saskatchewan's early expertise in space science (starting in the 1940s); some growth is due to the needs of the mining industry (instrumentation and control) and some is due to the research activities and needs of SaskTel.
67. There is no formal technology incubator at the University of Saskatchewan or at Innovation Place. UST has had an office in the park; it was not used and was closed.
68. http://www.innovationplace.com/html/frameset.html, January 2005.
69. Innovation Place, November 1997 fact sheet, and telephone interviews, summer 2002. In 1997, 64 per cent of the persons employed at Innovation Place worked for the private sector; 40 of the activities in the park were research, 24 per cent technical services, 33 per cent business services and 3 per cent education.
70. http://www.agwest.sk.ca/agwest.shtml. 'Our mandate at Ag-West Biotech is to initiate, promote and support the growth of Saskatchewan's agricultural biotechnology industries and the commercialization of related food and non-food technologies, by working with industry and external stakeholders…'
71. Number of head offices estimated from the Blue Book of Canadian Businesses, 1997, Canadian Newspaper Services International Ltd.
72. FP Markets (2001).
73. http://www.calgarytechnologies.com.
74. City of Calgary industrial base, 21 February, 2004, at http://content.calgary.ca/CCA/City+Hall/Business+Units/Customer+Service+and+Communications/Corporate+Marketing/Municipal+Handbook/+Welcome+to+Calgary/Industrial+Base.htm#researchpercent20andpercent20development.
75. Its debt load became unmanageable because of its rapid expansion and a business model which included extensive loans to systems customers (Langford et al., 2002, p. 3).
76. *The Globe and Mail*, 24 July, 2000.
77. Technology Enterprise Center (1993), p. 44: 'the incubator program provides affordable space with office services, on-site business counselling and access to a network of professional, financial and technological assistance'.
78. From interviews conducted in the summer of 2000.
79. The Entrepreneurship MBA program, specially designed for start-up entrepreneurs, was terminated in 2000; entrepreneurship and innovation is now one concentration in Calgary's MBA program.
80. From interviews conducted in the summer of 2000.
81. Ibid.

4. Mexico: the challenge to create regional innovative environments

INTRODUCTION

Mexico, as an industrializing nation, faces significant challenges in its efforts to develop successful innovative environments. This chapter begins with a review of the historical perspective that reveals the creation of scientific and technological capabilities as well as missed opportunities and numerous obstacles which have hampered the country in attaining its full potential, and shows that even those already acquired capabilities have disappeared in some cases. This is followed by an analysis of the country's existing national innovation system (NIS) which provides convincing evidence that technology-based firms (TBFs) and research centers (RCs) are the main agents for promoting innovation. These two types of institutions have ties to the educational system and the complementary hybrid organizations that function as sources of knowledge and expertise, and they serve as bridges to the production sector. In spite of their critical role, the public incentives that have been enacted to encourage and support these institutional elements have not been adequate to offset the financial risks inherent in supporting innovation. The science park and technology incubator mechanisms that were supported by government policies during the 1990s were not sustained, and as a consequence, by the year 2000, Mexico had only one science park and six incubators left, all in a relatively precarious state of operation.

This study has identified six knowledge regions or technology innovation poles (TIPs) that span several areas of the country. In order to facilitate comparisons of these six TIPs with the selected US and Canadian cases, they have been consolidated into four regional case studies: Case 1 is the Querétaro–Bajío region; Case 2 includes the medium-sized cities of Cuernavaca and Ensenada; Case 3 includes the large cities of Guadalajara and Monterey; and Case 4 covers Mexico City.

This chapter presents data about innovation poles on a broad regional level (somewhat similar to Chapter 3 for Canada), rather than more focused case studies of the main science parks and incubators of selected TIPS (as in Chapter 2 for the US). Each technology innovation pole is analyzed by looking at its research centers' (RCs) infrastructure and TBFs, as well as the knowledge linkages of which they are a part. These linkages are then examined at both a local and a national level, to determine how many of the pole's knowledge sources are internal or external to its surrounding region, or even international. Finally, a coefficient of technological specialization is calculated for each TIP, a measure of both the number of research projects in RCs and of the number of products in TBFs, by industry.

At the end of this chapter, an overview of the components of the innovation system is provided and a report on its scientific and technological infrastructure is presented. Overall, investment in the knowledge sector in Mexico tends to be cyclical, and there has been a decrease in the resources allocated to the technology sector of the country since 1995. There are some encouraging signs of a potential turnaround in this trend, but, by 2004, the new upswing in technology infrastructure and innovation support that was expected since 2001 had still not fully materialized. A correlation analysis shows the positive impact of incubation mechanisms – science parks or incubators – on the creation of innovative firms and between the number of TBFs and the number of innovations in each region. However, a lack of funds and absence of other government incentives that would otherwise accelerate the development and validation of new technologies for marketable innovations, make it unlikely that innovation activities will increase in Mexico in the near future. As a result, the current challenge is simply to sustain the weak regional innovation capabilities that are badly needed for Mexico to maximize the benefits derived from its NAFTA partnership and other worldwide economic relationships.

4.1 THE MEXICAN INNOVATION SYSTEM

This section lays out some key issues about the context of the innovation poles in Mexico. It starts with a historical perspective of the origins of Mexico's current innovation capabilities. It is focused on each region's educational and science and technology profiles, and their relationship to production. It then addresses the institutional framework and describes some components of the national innovation system (NIS), including its core capabilities and key institutions as well as the programs that are lacking. Finally, it analyses the location of innovation capabilities from a regional point of view as an introduction to the four innovation pole cases.

4.1.1 Socio-economic, Science and Technology Overview

Mexico is the 11th most populated country in the world, with around 100 million inhabitants[1] and an average annual population growth of 2.2 per cent in the early 1990s and 1.6 per cent from 1995 to 2000. One quarter of the country's population lives in small towns with less than 2500 people and another quarter (26.3 per cent) lives in big cities with more than half a million inhabitants. The remaining half lives either in semi-rural areas (13.7 per cent in towns of 2500 to 15 000 inhabitants), or in small and medium-sized cities (14 per cent in cities of less than 100 000 inhabitants and 21 per cent in cities of 100 000 to 500 000 inhabitants respectively).

Nearly 40 million people (12 years of age or more) are economically active. Yet only 7.9 per cent of total employment is held by tertiary-level graduates,[2] and there is only one person per 1000 employees engaged in R&D activities.[3] School enrollment is 3.4 million in kindergarten, 14.8 million in primary school, 5.1 million in secondary, 2.9 million in preparatory school (thus a total of 8 million are in secondary level) and 2.0 million in universities or higher education institutions (Reséndiz, 2000, p. 95). The statistics show that 10.6 per cent[4] of the adult population (over 15 years old) is illiterate and only 93 per cent of children (6–14 years old) attend school, which illustrates well the polarization in Mexican society.

The Gross National Product in 2000 was $574 billion, divided into 4.3 per cent in agriculture, 28 per cent in industry (of which 73 per cent are in manufacturing), and 67.7 per cent in services (INEGI). For 2005, the expected GDP is $669 billion for 106 million inhabitants, or $6300 per head.[5] Mexico is ranked 13th in the world for its exports, which account for $166 billion or 2.6 per cent of the world's exports. It imported $177 billion in 2000.[6] 'Maquiladora'[7] industries account for 47 per cent of exports and 41 per cent of imports, and oil makes up 10 per cent of exports, with a production of 2.9 million barrels per day.

A short historical review of applied knowledge and know-how in Mexico shows two lines of development: on the one hand, technical capabilities have emerged historically out of indigenous specialities in medicine, astronomy, construction and architecture; and on the other, foreign knowledge has been incorporated in industries during certain periods in a way that has prompted spurts of economic growth but without a sustained long-term development.

Together, these two processes have given rise to an *extensive* type of production, based mainly on labor that contrasts with *intensive* types of production, based on the use of technology. Historically, extensive production traces its roots to the Spanish Conquest in 1521, when the conquerors pillaged indigenous communities and imposed European economic practices that disrupted local traditions and brutally subjugated the natives. In this context, there were few incentives to innovate, as natives were concerned mostly with their own immediate survival and keeping their cultural practices, and Spaniards had a seemingly inexhaustible supply of low-cost workers at their disposal.

An important change came under Carlos III, as the Borbonic Reforms of 1763 promoted the use of new methods and manufacturing techniques in the sugar, tobacco, textile and coinage industries. As a consequence of these reforms, the first scientific institutions were created, including in particular the Royal Mining Seminar, founded in 1792.

The demand for new technical knowledge increased further, on one hand due to the Industrial Revolution (1850–1934), and on the other as a result of the foundation of technological and scientific institutions dealing with geology, medicine and astronomy.

Yet these processes had mixed results. They increased production and expanded it horizontally, but they failed to stimulate vertical integration of local industries, many of which continued to import rather than develop certain inputs, such as machinery and industrial products. The industrial revolution also increased the wealth of a few rich and powerful families without generating a better income distribution, as workers or *peons* were often paid in goods rather than wages in the *haciendas* (*lista de raya*).[8]

From 1935 to 1995, post-revolutionary governments in Mexico implemented industrial policies aimed at filling those productive gaps, but the partial results they obtained did not solve the basic economic problems mentioned. As a result, from the point of view of industrial capabilities, these periods can be called *truncated modernizations*.

Three such periods can be identified. The first one (1935–1970), saw improvements in innovation and research capabilities with the creation of the National Energy Research Centers in oil, electricity and nuclear areas. Also in this period, a coordination agency was founded for science and technology policy (Conesic, which later changed to INIC, National Institute for Scientific Research) (Casas, 1985).

The second period of modernization began with the founding of CONACYT (National Science and Technology Council) in December 1970, which promoted the creation of research centers all over the country and raised the number of scholarships for graduate students within the country and for study abroad. During this period, the Mexican government paid special attention to universities and technological institutions. However, due to the Mexican economic crisis of 1982 public funding for higher education was significantly reduced, impacting laboratory acquisitions and equipment maintenance and imposing reduced salaries for professors and scientists. The university crisis was partially solved by the creation of competitive funds based on scientists' and academics' productivity indicators as a new National Research System (NRS) was established in 1984.[9]

The third period of modernization began in 1994 with NAFTA (North American Free Trade Agreement) as part of a new open economic model. The outburst of guerrilla activity in Chiapas at this time focused attention on some demands of the indigenous population and served as a reminder that Mexican society continues to be polarized on the basis of education as well as income levels, in a way analogous to that established at the time of Spanish Conquest.

Currently 'maquiladoras' dominate the landscape of Mexican production. The result is that Mexico's economy continues to face truncated modernizations as the maquiladoras make extensive use of labor or are limited to the simple mechanization of traditional industries, or the use of modern technologies without enough efforts to build local innovative capacity. However, to achieve the goal of having a modern innovation system, Mexico needs to generate a core of its own technological capabilities instead of merely using imported modern technologies. In this respect, some of the main issues to focus on are to promote the development of TBFs that will aid in the transfer of technologies from local research centers.

4.1.2 The National Innovation System (NIS)

In order to understand some recent institutional changes in Mexico, it is useful to consider its transformation from a closed market economy with an import substitution model to an open market economy. This process began with Mexico's entrance into the GATT (now the WTO) in 1986, and since 1994 it has continued with a series of international trade agreements, mainly NAFTA.

In this global economic environment, Mexican organizations can be grouped in terms of the way in which they contribute to innovation:[10]

- **TBF** and **research centers** are involved directly with the innovation process.
- **Education and training** develop human skills and involve organizations that participate in and promote the flow of science and technology.
- **Financing and incentive** provide organizations the ability to stimulate innovation.
- **Public policy,** regulations and programs promote or impede the creation and operation of knowledge and innovation environments.

TBFs and research centres

The study of TBFs in Mexico is very recent. It is estimated that there are around 1000 TBFs, half of which were created in the 1990s (Corona, 1994). There are other types of firms that have been innovating but the strategic changes they make tend to be aimed at modernizing their technology processes (technological change). As such, these firms *use* new technologies from external sources, instead of working with those sources to develop their own capabilities to generate new technology.[11] Industry modernization, then, involves adapting new technology to their existing industrial needs or developing only certain technologies on a restricted basis. Yet the use of new products or processes does not build up and sustain technological capabilities. Rather, it is the generation and implementation of new products and processes achieved by TBFs that is a better indicator of the country's technology capabilities. Thus TBFs are main agents of innovation, while research centers and other firms' technology providers are primary agents of knowledge production.[12]

Education and training

Education contributes to a positive innovation environment, and higher education in particular plays an important role in developing scientific and technological capabilities. Yet in order to fulfill this role, the Mexican higher education system needs to address some barriers to research activities experienced mainly in its graduate programs.[13]

Recently, CONACYT has initiated a program called 'Provinc' that focuses on the creation and promotion of technology management units within research centers and universities (UGCT) in order to provide training to build capabilities in technology transfer and production methods.

Financing and incentives

One of Mexico's main problems is the virtual absence of venture capital to finance innovation. Some public programs have tried to address this problem, but they have not been entirely successful. CONACYT and NAFINSA manage the FIDETEC (Fund for Research and Development for Technological Modernization), founded in 1991 to finance pre-competitive R&D, technology transfer and product and processes innovation. FIDETEC has had a very low impact on innovation, in part because of procedural problems linked to its coordination by commercial banks whose personnel were not trained to understand investments in innovations. Nevertheless, in coordination with the PIEBT (CONACYT's program for funding technology incubators), FIDETEC stimulated the creation of TBFs during the 1990s.

Starting in 2002, innovation firms received a fiscal credit incentive equivalent to 30 per cent of their spending on research and development to improve products and processes,[14] and in 2003, seven start-ups were selected to receive funding (1.8 million dollars in total) from a new CONACYT program (AVANCE).

Public policy

The Mexican government has kept a low profile regarding innovation, lagging behind other industrializing countries. The main institution promoting and coordinating public policy for science, technology and innovation is CONACYT. Since 2002, the General Council for Science and Technology (CGCyT)[15] has brought together all of the public agencies involved in these matters.[16] Moreover, the coordination on a national level of the states' science and technology councils, as well as the development of a Consultative Forum and networking among public research centers, have the potential to stimulate a new wave of science and technology activities.[17] In order to be successful, this top-down organization will require the development of R&D networking based on decisions at grassroots level that will stimulate the creation of a positive innovation environment including improvements in sources of funding to offset the risks associated with innovation activities.

The National Science and Technology Program (Pecyt, 2001) proposed several ambitious changes to be implemented during the current presidential period 2001–2006 (see Table 4.1):

- To increase its R&D funding from 0.4 per cent to 1.0 per cent of GDP.
- To transfer a larger amount of these resources to the states.
- To increase the share of national R&D carried out by science and technology firms from 23 per cent (in 2000) to 40 per cent.

Unfortunately, by 2003, R&D support continued to account for only 0.4 per cent of GDP, and firms contributed only 30 per cent of national spending on R&D. Yet in spite of its failure to attain specific goals, Mexico has enacted some important institutional changes since 2001 that have helped to increase the amount of national resources dedicated to R&D.[18]

4.1.3 The Regional (State) Systems of Innovation

In 1995, CONACYT created the Regional Research Programs,[19] by regrouping states into nine regions in 1997 and 11 by 2002 after a reclassification of certain regions (Figure 4.1). Through these programs, the SIGHO region (where the Querétaro–Bajío poles are located) saw a 22 per cent increase in the proportion of its researchers included in the national system (NRS), from 5.4 per cent in 1996 to 6.6 per cent three years later. Yet other regional research systems (SIMORELOS, SISIERRA, SIVILLA, SIZA) saw few improvements, and still others experienced decline; SIMAC declined by 2.4 percentage points and the Central Zone declined by 2 points. Apart from SIGHO, members of the National Research System (NRS) saw no clear improvements in regional participation (see Table 4.2).

By 2004, aside from regions that were again reclassified, the main programs were those involved with state funding (called 'Fondos Mixtos'[20]). These 'mixed' funds were developed in partnership with the state governments,

Table 4.1 Mexico: main science and technology indicators

Indicators	2001	Goal 2006
Gross Domestic Expenditure on R&D – GERD, as a percentage of GDP	0.4%	1.0%
Percentage of GERD financed by industry	26%	40%
Funding for priority sectors research	$75 m*	$2676 m*
Mixed funding for regional–state research	$10.7 m*	$535.2 m*
Total R&D personnel per thousand active population	0.7	2.0
Total business enterprise R&D personnel as a percentage of national total	20%	40%
New researcher's posts in the Public Research Center	60	12500**
New researcher's posts in higher education R&D	120	15500**
Total Federal Government Budget for R&D	2%	4.0%

Notes:
* Millions of 2001 dollars (9.34 pesos/dollar)
** For the whole period 2001–2006.

Source: Pecyt (2001)

Table 4.2 Regional research systems: national researchers and regional gross domestic product

Regional Research Systems (RRS)	1996		1999		Gross Domestic Product	
Conacyt(1)	Researchers Number	Researchers Per cent	Researchers Number	Researchers Per cent	1996 Per cent	1999 Per cent
Metropolitan*	3899	65.3	4584	63.2	34.8	34.5
South: SIBEJ	69	1.2	90	1.2	5.2	4.9
Centre: SIGHO	321	5.4	477	6.6	7.8	7.8
Orient-Golf: SIGOLFO	108	1.8	136	1.9	5.9	5.5
Peninsula, Northwest: SIMAC***	533	8.9	470	6.5	8.5	8.4
Occident: SIMORELOS	343	5.7	467	6.4	9.9	10.2
Northeast: SIREYES	242	4.1	298	4.1	12.4	13
Southeast: SISIERRA	109	1.8	161	2.2	3.8	3.7
North-Centre: SIVILLA	68	1.1	100	1.4	6.2	6.5

Orient-Centre: SIZA	277	4.6	367	5.1	5.3	5.4
Other**	0	0.0	102	1.4	0	0
Total	5969	100	7252	100	99.8	99.9

Notes:
(1) RRS, with the new regions' names since 2002 and the former ones in capital letters. See Figure 4.1.
* This Region has been a Conacyt RRS since 2002: DF, State of Mexico and Morelos.
** Other: Foreign Institutions and others not regionally located.
*** Since 2002, SIMAC region has been divided in two: North-west (Sonora and Sinaloa) and Peninsula of Baja California.

Source: Own elaboration based on CONACYT, Statistical Data (1990–99).

which financed 43 per cent of the $14 million raised in 2001; 45 per cent of $22 million in 2002, and 40 per cent of $35 million in 2003.

In fact, the ten federal states that are home to the selected regional cases (see section 4.2 below) account for around 80 per cent of NRS members (81 per cent in 1999, 78 per cent in 2003). More than half of these researchers are found in the Mexico City metropolitan area alone, which includes the Federal District (DF) and the State of Mexico (57 per cent in 1999, 53 per cent in 2003). Of the other cases, each account for 3 per cent (Baja California State) to 7 per cent (Jalisco and Michoacan States) of national researchers.[21]

In order to assess the impact of CONACYT's regional programs for innovation, it is necessary to look for other indicators, such as the participation of high-tech industries in the performance and financing of R&D.[22]

Conclusions

Mexico's innovation capabilities have been relatively poorly developed as its dominant industries have traditionally been based on labor-intensive production processes, which is nowadays the main characteristic of 'maquiladora' industries. Breaking with the past, the country's current open economy model must be reoriented to promote intensive technology industries above and beyond the industry modernization. This can be done by exploiting the advantages that come from participating in international economic agreements, such as NAFTA.

How to do this will depend on how well the risks of innovation are counterbalanced by the impact of recent institutional changes in Mexico, including in particular the coordination of public agencies and state R&D efforts.

4.2 THE SELECTED MEXICAN CASES OF REGIONAL INNOVATION POLES

Technology-based firms, research centers, and incubators and science parks are concentrated mainly in six regions.[23] These six regional innovative poles have been grouped into four cases according to their population size[24] and other variables explained below.

Case 1, Querétaro–Bajío is a large region with a substantial number of technology research centers; Case 2, Ensenada and Cuernavaca, consists of two medium-sized cities with active research centers that promote technology transfer and innovations; Case 3, Guadalajara and Monterrey, two big cities with a large industrial base and programs for the creation of innovation capabilities based on new technologies; and Case 4, the metropolitan area of Mexico City with a large number and variety of education facilities, research centers and some innovation firms, in many industries (Corona, 2001).

In each of the selected TIPs, TBFs and research centers have been the focus of the study. Overall, 262 TBFs have been surveyed in three time periods. The first two periods, both named INDICO (Innovation, Diffusion and Competitiveness) occurred respectively in 1994–1995 (Indico1) and in 1996 (Indico2). This last study was aimed to assess the impact of the 1995 Mexican economic crisis on innovation activities.[25] The third period, called Monarca, occurred in 1997–1999, when began the comparative study of NAFTA countries, the US, Canada and Mexico, when 66 research centers were visited.[26]

As a consequence of this work, it has been decided to focus our analysis on six innovation poles regrouped into four cases. Two of them are innovation regions: (1) the Querétaro-Bajío has the highest national concentration of technology research centers and (2) Mexico City, the largest metropolitan area, includes several knowledge zones built around research centers and universities. The other two cases, each including two innovation poles are: (3) the medium-sized cities of Ensenada and Cuernavaca, and (4) the large industrial cities of Monterrey and Guadalajara, both having good examples of TBFs and innovation mechanisms for study. The rest of this chapter deals with these four cases (Figure 4.1).

4.2.1 Querétaro–Bajío

The Querétaro–Bajío (Q–B) region, located in the center of Mexico (Figure 4.1), is an industrial corridor that extends more or less continuously from Querétaro to the Bajío in the state of Guanajuato (population is 1 million in the industrial corridor of Querétaro and 2.3 million in the Bajío region). These regions make up 50 per cent and 72 per cent respectively of their state's populations. Each region has one incubator, in different stages of development, and only one science park has been built in Querétaro since 1991.

The Querétaro industrial corridor is situated at the crossroads between Mexico City and the North of the country. The main industries are auto parts, textiles, electric appliances and electronics devices, food, chemistry and paper. This industrial profile evolved over time from the traditional food and textile industries (1940–60) to the large modern steel and automotive sectors and multinational firms in food and chemicals (1970s).

The Bajío is a traditional region based on agriculture whose industrial corridor developed when the petrochemical plant of Salamanca started operations in 1950.

Regional technology infrastructure

Both states have a good educational infrastructure. The main universities are

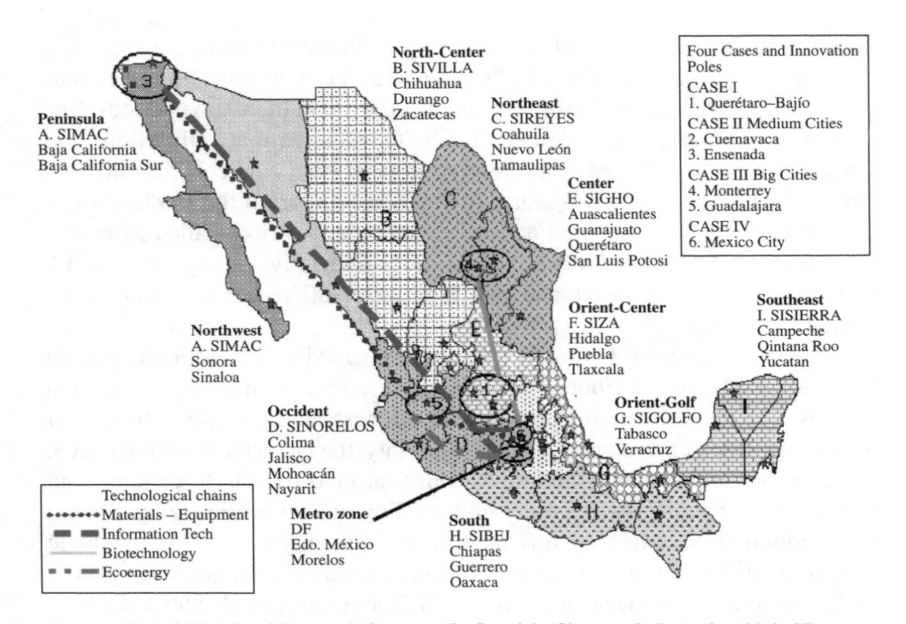

Notes: The old Regional Research Systems (in Spanish 'Sistema de Investigación', SI) names are:

SIBEJ, Benito Juárez	SIMAC, Mar de Cortés	SISIERRA, Justo Sierra
SIGHO, Miguel Hidalgo	SIMORELOS, José María Morelos	SIVILLA, Francisco Villa
SIGOLFO, Golfo de México	SIREYES, Alfonso Reyes	SIZA, Ignacio Zaragoza

Figure 4.1 Mexico: innovation poles and technological chains

the University of Querétaro (UAQ) and the University of Guanajuato (UGto), with faculties and schools located in different Bajío cities (for example FIMEE in Salamanca, which offers programs in mechanical, electrical and electronic engineering).

Both regions have improved their technology infrastructure. By 1999, they added 20 research centers – 12 in Querétaro, and eight in Bajío – with 2700 employees, including six public research centers established by CONACYT, five national government labs, five university research centers and four research units within large firms, all of these in Querétaro.[27]

Most of the public research centers are medium in size[28] (making up 45 per cent of the total) and they were founded during the presidential periods of 1976–82 (Lopez-Portillo) and 1988–94 (de la Madrid). The six RCs founded during the first period seem to have been built with relatively abundant public resources available at that time (obtained from the oil economy), while the seven RCs founded during the second period are better understood as the outcome of a deliberate process of decentralization out of Mexico City, after the earthquake of 1985.

The knowledge infrastructure is based on research institutions established both within the region and outside it. In order to explore this distribution, a question was asked to each of the main research centers (in the region) to rank their relationships to other research centers and universities. The result shows a roughly equal division between internal and external institutional relationships for the region as a whole (25 internal, 27 external). Querétaro is mainly interconnected with local institutions (78 per cent, 11 local links out of 14) and its external links are mostly with RCs located in Mexico City; while the Bajío pole's RCs are mostly interconnected with outside institutions (63 per cent, 24 links out of 38) mainly with Mexico City (10 links out of 12), and with international RCs (30 per cent of the 24 external links).

Overall, Q–B is characterized by strong links between knowledge institutions: first, located both within the pole; second, in Mexico City and; third, abroad. And as noted above, most of the external links originate in the Bajío, while those in Querétaro are primarily local.

Both poles have an infrastructure of research centers that is larger than the national average and that supports a positive environment for innovation.

Incubation mechanisms

The incubator in Querétaro (PIEQ) was founded in 1992, and the one in Bajío (CENIT) was founded in 1994. The Querétaro incubator was originally located in a rented house, where it hosted six TBFs. In 1994, it moved to a new building located in San Fandila, a high-tech land development 11 kilometers south of the city of Querétaro, which also hosts the Bernardo Quintana Industrial Park housing the National Standards Center (CENAM), and the National Institute of Transportation (IMT). After this move, the PIEQ incubator became part of a technology park and, between 1994 and 1999, it hosted a maximum of 11 firms. Cash flow problems caused it to declare bankruptcy in 1999, along with the San Fandila Science Park which hosted three research centers but no firms. Over its five-year period of operation, this incubator hosted 25 firms with three graduations.[29]

Bajío's Center for Business and Technological Incubation (CENIT) was founded in 1994 and operated as a take-off project from 1996 to 2004. It is located in the Delta industrial park of the city of Leon and is a joint project headed by two local research centers (CIO and CIATEC).[30] This incubator is more technology-oriented than PIEQ, as expressed by its selection criteria for firms, which must be: 1) a business technology spin-off, 2) a research-entrepreneur moving from prototype stage through process development, tech-transfer and scale-up, 3) an entrepreneur looking to develop a product or service to meet a perceived market need. From the point of view of money flow, CENIT's purpose is to do business and make a profit,[31] and its policy is to participate as a shareholder in up to a third of the firm's private equity.

In the first three years, TBFs were attracted into the CENIT incubator by the CONACYT Incubators Program and the possibility of accessing venture capital from the NAFIN-CONACYT Fund (FIDETEC). When these government programs did not perform as hoped, many innovative firms lost interest. The incubator's failure has been partially blamed on poor management. It did not address TBFs managerial consulting or funding needs and failed to cultivate institutional knowledge links to support TBFs' in generic or specific technologies. The 1995 economic crash dealt the final blow, as venture funding diminished and innovation support withered away (see note 25).

There were four incubating firms[32] in 1998 and by 2001 no firms were left. In an effort to survive, the Bajío incubator (CENIT) sought to move into the University of Guanajuato's liaison office (INCU-VEN), where by 2005 it is taking off.

TBFs surveyed

Eighty-seven innovative firms were selected and surveyed, mostly outside the incubators in the Querétaro (68 TBF) and Bajío (19 TBF) poles. Most are micro and small firms (65 per cent), with a much smaller number of medium (18 per cent) and large (16 per cent) firms.[33] This correlates well with the findings of other studies, and makes sense in light of the fact that the spread of TBFs in Mexico is a recent phenomenon (Corona, 1997).

TBFs in both the Querétaro and the Bajío regions show similar patterns in terms of their knowledge links. Their main sources of technology are other firms, research centers and universities, and they have more links with firms located outside the pole than within it, but the relationships they develop with research centers and universities located within the pole are more intensive. This pattern can be identified as a 'positive knowledge pole'.

The TBFs of Querétaro–Bajío are involved in a variety of technology areas, with the main focus on machinery and equipment (24 per cent), transportation equipment (20 per cent) and biotechnology (20 per cent). Some differences between Querétaro and Bajío include the greater proportion of agriculture and transportation equipment firms in the Bajío and a greater emphasis on machinery and equipment (26 per cent) and electronics and software (7 per cent) in Querétaro.

TBFs in the region are among the most innovative in the country, especially in Querétaro, with a very high correlation between number of firms and number of innovation by industry (Table 4.3).

In order to compare the firms' innovativeness, an average number of

Table 4.3 Correlation between the number of innovations by industry and the number of firms (TBF)

Inno (i)= C+b (TBF)i i= industry	Constant $(TBF)_i$		R^2
Innovation of industry i in the region of:	C	b	
B (Bajío)	0.10	3.50	0.83
Q (Querétaro)	3.47	3.08	0.89
Q+B	3.37	3.21	0.92
Cuernavaca	-0.08	2.91	0.95
Ensenada	1.18	2.75	0.81
Guadalajara	0.23	2.31	0.67
Monterrey	0.02	4.12	0.96
Mexico City	1.14	4.42	0.76
MEXICO	4.96	3.32	0.86

Notes: Number of observations 15, according to the number of industries.

C = constant
b = correlation coefficient
R^2 = determination coefficient

Table 4.4 *Comparative assessment of the technology innovation poles (cases)*

Case	Tech pole	RC*	Direct Incubation Mechanisms (DIM)**		Innovation					
					Average performance for region***		Milieu	Specialization index for local RCs****		
			Incubators	Science Park	Number of TBFs	Average performance per TBF***		Tech areas	RC	TBF
I	Querétaro Q	8	PIEQ (K)	Fandila (K)	68	3.1	3.5	Telecommunications	2.9	
								Transp Equipment	2.9	−0.3
								Services	2.8	−0.3
								New Materials	1.3	
								Machinery	0.9	1.3
								Electronics	1.6	−0.4
								Total	12.4	0.3
						3.2 (QB)	3.4 (QB)			
	Bajío B	12	CENIT (S)	—	19	3.5	0.1	Agriculture	4.1	1.5
								Electronics	1.6	
								Biotechnology	1.5	0.4
								Software	1.3	
								Machinery	1.0	0.3
								Electronics	1.6	
								Transp Equipment		1.8
								Total	11.1	4.0
	Ensenada	4	IEBT (K)	—	15	2.7	1.2	Telecommunications	9.2	3.7
								Instruments	1.0	7.3
								Electronics	0.9	1.2
								Total	10.2	11

II										
	Cuernavaca	6	CEMIT (K) IETEC-ITESM (K)	TPM (P)	40	2.9	−0.1	Biotechnology	4.8	1.5
								Software	−0.8	2.5
								Energy	0.1	1.1
								Total	4.1	5.1
	Monterrey	7	Entrepreneurship (+) (ITESM)	—	14	4.2	0.0	Ecology	2.7	
								Software	1.1	
								Total	3.8	
								Chemistry		3
								Biotechnology		2.6
								Energy		0.7
								Total		6.3
III										
	Guadalajara	5	CUNITEC (+)	Belenes (+)	32	2.7	0.2	Machinery & Eq	4	
								Agriculture	5.5	
								Electronics	1.5	
								Total	11	
								Ecology	0.4	3.5
								Telecommunications	−0.3	1.2
								Energy		1.2
								Biotechnology	0.6	0.5
								Total	0.7	6.4
IV	Mexico City	13	SIECYT-UNAM (K) Cetaf-UACH (P) Torre Ing-UNAM (K, S) CIEBT-IPN (+) UAEM-Tecamac (N)	Ferrerias (P)	55	4.4	1.1	Pharmaceutics	12.8	1.5
								Instruments	10.8	0.5
								Chemistry	11.7	0.2
								Energy	9.1	0.3
								Ecology	2.8	−0.1
								Total	44.5	2.4

National (or Best)	66	2 TBInc (+)	Belenes (+)	243		3.4	Total		97.6	29.2

Notes:
* RC, Research Center
** State of the Direct Incubation Mechanisms (DIM): Planned (P), Bankrupt (K), Surviving (S), Operating (+). Technology Incubator unless indicated as non-technology (N)
*** QB: Querétaro and Bajío regions taken together
**** The specialization coefficient is calculated as follows:

$$S_i^j = \frac{\text{Industry } (i) \text{ of Pole } (j) \text{ / All Industries of Pole } (j)}{\text{Industry (i) National / All Industries National}}$$

It is the share of industry i in each Pole j, divided by the National share of that industry.
The coefficient is normalized, giving zero to the national specialization (rather than 1, (SP (j) = S_Pole (j) – 1).

innovations per industry is calculated. Bajío's agricultural sector is the 'most innovative industry' in this respect, with 6.7 innovations per firm.[34] Averaging across all industries, the Q–B Region has an average of 3.2 innovations per firm, and, given the constant value of 3.4 innovations, it is expected to have a permanent positive environment for innovation (Table 4.4). Innovation would therefore be stimulated if both poles, Querétaro and Bajío, opened as a single knowledge region, increased integration creating more opportunities for networking.

TBFs in both poles have similar industrial patterns, and they have links to research centers with similar project technology areas (PTA). These are materials and equipment (46 per cent TBF and 21 per cent PTA) and biotechnology (20 per cent TBF and 21 per cent PTA). Yet the research centers' projects focus mostly on services (31 per cent), secondly on biotechnology (10 per cent) and thirdly on new materials (11 per cent). This suggests that biotech research projects are associated with innovations in agricultural firms (15 per cent), and the 'services' research projects are commissioned by many more firms than the 7 per cent characterized as services firms.

It becomes interesting to compare the innovations of TBFs relative to research center projects within each industry.[35] The Q–B region is remarkably strong in most of the technology areas examined as shown by its specialization coefficient.[36]

First, the relative contribution in each technology area of research projects reported by local research centers is measured against the national parameters. According to this indicator, Querétaro is strong in telecommunications (2.9), new materials (1.3), transportation equipment (2.9) and services (2.8); the Bajío region is strong in software (1.3), biotechnology (1.5) and mainly agriculture (4.1), and it is near the national average for firms in instruments (−0.1). Both regions have specialties in electronics (1.6), machinery and equipment (0.9, 1.0, respectively). The region's weak areas are energy and ecology (Table 4.4).

Second, the relative contribution of TBFs in each technology area compared with the national average is estimated. According to this measure, the TBFs in Querétaro are strong in machinery (1.3) and those in Bajío are strong in transportation equipment (1.8) and agriculture (1.5). Both areas are near the national average for firms in chemistry (0.1) (Table 4.4).

In order to better exploit this R&D infrastructure, it will be necessary to develop a policy of partnerships between industry and research institutions, so as to share in the risks and benefits of innovation. Querétaro–Bajío region's incubation and science park failures suggest that deliberate steps must be taken to enhance the innovation environment. State and local governments have a crucial role to play, to promote and even participate in this process.

Conclusions

The Querétaro and Bajío region as a whole offers a positive environment for innovation, with a good infrastructure of research centers that is among the largest in the country. Both of the region's poles are located in central Mexico and have the following kinds of innovation sources: 1) application of new technologies and new organizational procedures to traditional industries, including machinery and transportation equipment and agriculture, and 2) new innovation firms or TBFs, which expanded from either traditional or new technologies to generate specific innovations in biotechnology, electronics and software.

However, it is apparent that the good research infrastructure of the Q–B region is not being properly exploited for innovation. This problem needs to be addressed by a policy facilitating the entrance and creation of more innovative firms in the region, mainly in new technologies (electronics, telecommunications, new materials, biotechnology and software). One current obstacle is the weakness of incubation mechanisms for supporting innovation and the absence of venture capital funds.

Dedicated incubation support mechanisms – incubators and science parks – have failed in both Querétaro and Bajío. A possible explanation for this is the shortage of financial resources, which made it impossible for them to be sustained over a period long enough to reach a self-financing stage. Complementary public funds were too scarce, either at the stage of technology feasibility or for the development and introduction of innovations on the market.

In conclusion, the sustainability of the Querétaro–Bajío as an integrated innovation region depends on the recognition that the region's two poles have many complementarities in their knowledge infrastructure and their basic, applied or market-oriented research. In addition, Bajío shows a 'positive' sign in the sense that it has a good balance between inner and outer sources of knowledge. As a result, an influx of venture capital to the region would likely raise the number of innovative firms and promote a better diffusion of innovation in services, products or processes, or all of them.

4.2.2 Medium-sized Cities: Cuernavaca and Ensenada

Cuernavaca and Ensenada are medium-sized cities with more than 300 000 inhabitants (370 000 in Ensenada and 339 000 in Cuernavaca, in 2000). Cuernavaca is the capital of the state of Morelos, located in the center of Mexico, 60 kilometers south of Mexico City and Ensenada is located 60 kilometers south of the US–California border in the state of Baja California, in the northwest of Mexico. These two TIPs are separated by 1400 miles, but they have some innovation similarities. First, both cities have established a process for TBF creation in which a local research center is the main supporting institutional actor. In Cuernavaca, this champion institution is the National Electrical Institute (IIE), founded in 1975, and in Ensenada it is the CICESE, founded in 1973 and oriented to telecommunications, physics and oceanic research. In both cases, the creation of a specialized knowledge environment was triggered by the foundation of these research institutions that attracted senior researchers from Mexico City.

Regional Technology Infrastructure

Cuernavaca attracted some important research centers, in part because of its proximity (60 kilometers) to Mexico City's large and diversified research infrastructure. It hosts some important national research labs, including the above-mentioned National Electrical Institute (IIE, founded in 1975) and the National Institute for Water Resources (IMTA, founded in 1986). It also hosts some research centers of the National University (UNAM), including biotechnology (IB, 1982), physical sciences (1982, CCF 1998), and energy (CIE – UNAM, 1996).

The Ensenada pole includes the nearby cities of Tijuana and Mexicali, capital of the state of Baja California, both of which have numerous 'maquiladora' industries and share borders with the US. This large innovation region encompasses 94 per cent of the state's population. Ensenada was traditionally involved in the fishing and canned fish industries, and the settlement of research centers there encouraged a shift to new technologies. Ensenada's CICESE research center, oriented to oceanography and electronics, is complemented by the IPN's electronics and automation center (CITEDI, located in Tijuana since 1982) and the physics research center of the UNAM (CECIMAC, since 1997). This region's proximity to California, with its US universities and research institutes (such as the Scripps Institute of Oceanography, La Jolla, CA), has stimulated the research and high-tech entrepreneurship environment.

Ensenada has only half the research center infrastructure present in Cuernavaca, but both poles' RCs employ around half of the NRS members in their respective states. Moreover, Cuernavaca and Ensenada have research centers that are polarized in size. They have respectively three and one large research centers (more than 250 employees), no medium sized ones, and they each have three small ones (less than 100 employees). Local research centers in both regions reported having links mainly with research centers and universities in Guadalajara and Mexico City, as well as with other Mexican poles, but they did not report any links to Bajío and Monterrey. An information innovation chain can be traced from the northwest to Guadalajara and Mexico City (Figure 4.1).

Cuernavaca maintains more links to external than internal research institutions (ratio: 1.5), while Ensenada has an almost equal number of links to research institutions within and outside its innovation pole (ratio: 0.9). Cuernavaca's greater linkage to local and international institutions gives it a better knowledge link position overall.

Also, Cuernavaca is serviced and supported by Mexico City's research infrastructure and Ensenada is partially supported by some of the US research centers in California. These specialized technology regions count on leveraging the research areas outside of their poles to attract innovative firms. This regional pattern can be identified as 'satellite' innovation poles.

Incubation mechanisms

Both technology innovation poles have pioneering TBFs and incubators.[37] In Cuernavaca, the Centre for Technological Innovation Businesses (CEMIT) was founded in 1990 with five TBFs. It is part of a larger project, and the incubator was meant to serve as a first step toward building a Technology Park.[38] The fund created for this purpose had large organizational stakeholder support, which unfortunately resulted in a heavy decision-making process that obstructed incubator activities instead of giving them the desired leverage. The CEMIT was therefore located in CIVAC, the older and main industrial park of the state of Morelos built in the 1960s.

Cuernavaca's second incubator, the Incubator for Businesses in Technological and Administrative Innovation (IETEC) was founded in 1994 as an initiative of the local Morelos ITESM campus. It was financed by NAFIN (25 per cent), CONACYT (25 per cent) and the ITESM (50 per cent), it rented a building with room for 20 firms, and by 1998 was hosting 11 tenants, five of which were external. These tenants received technical services such as design, legal advice and business development counselling, as well as assistance with financial applications, mainly from FIDETEC. A four-month pre-incubation phase was allowed for elaborating the firm's business plan. In addition, they had access to an industrial manufacturing laboratory which they could rent on an hourly basis for product design engineering and production processing using numerically controlled equipment. The incubator closed in 1998, as operating costs were higher than revenues, and no other financial support was available.

In Ensenada, the Incubator for Technology-based Businesses (IEBT) operated in a rented building from February 1991 to January 1998.[39] It was supported by the CICESE Research Centre and five out of eight initial tenant firms emerged out of CICESE's research projects. Its full hosting capacity was 20 firms and it reached a maximum 70 per cent rate of occupancy. Many of the new firms were financed by the two CONACYT funds: the former Risk Share Fund ('Riesgo Compartido') and FIDETEC.

The stakeholders in the incubator were CICESE, CONACYT and Nafinsa (the latter two run FIDETEC). The criteria for firm entrance were more or less standard: technology merits, entrepreneurship abilities, quality of the product or service, market potential, financial status, and business plan.[40] Many of the bankruptcies that occurred were due to shortages of funds for the startup tenants that also impacted incubator operations as well as the impact of the 1995 Mexican economic crisis (see note 25).

TBFs surveyed

Fifty-five TBFs were selected from both inside and outside the incubators, for direct interviews – 40 in Cuernavaca and 15 in Ensenada. Most of these TBFs are micro and small in size (89 per cent) and only 11 per cent are medium-sized and large firms. Moreover, most are nationally owned, as only around 10 per cent have foreign capital.

The main sources of technology for these TBFs are other suppliers and consulting firms. In both cities, these sources are more frequently external than internal to the pole. The secondary sources of technology are other research centers and universities. In Cuernavaca, these are more often found locally (within the pole) than otherwise, while in Ensenada they are mainly found externally rather than internally. Overall, these TIPs can be characterized as mainly 'externally supported' with outside knowledge as source of innovation.

The largest number of innovations per firm occurs in Ensenada's agricultural sector and in Cuernavaca's ecology sector.[41] Overall, across all industries, a simple correlation analysis shows 2.91 innovations per firm in Cuernavaca and 2.75 in Ensenada. In the latter, there is a constant regional synergy of 1.18 innovations per firm, compared with a negative value in Cuernavaca (−0.08) (Table 4.3).

The two poles have different research technology specialties. Cuernavaca is quite strong in biotechnology, while Ensenada has expertise in telecommunications. Yet both poles fail to adequately link their regional research expertise to the TBF's specialties.

Cuernavaca's main research specialties are in biotechnology (4.8), while local high-tech firms are in software (2.5), biotechnology (1.5) and energy (1.1). Ensenada, which specializes mainly in telecommunications research (9.2), matches its research strengths partially with some TBFs (3.7), but less so with TBFs in instruments (7.3) and electronics (1.2) (Table 4.4).

In sum, a key competitive advantage for both innovation poles is the proximity to well-equipped and specialized research, as well as education infrastructure: Mexico City for Cuernavaca and California, US, for Ensenada. Each region has active research centers, which have promoted technology transfers and spin-offs by some researchers – IIE for Cuernavaca and CICESE for Ensenada. This trend peaked during the 1990s, when CONACYT promoted the creation of incubators. As a result, both of these regions played a pioneering role in developing technology incubators in Mexico.

Both innovation poles suffer from a lack of venture capital and the absence of policy mechanisms that would promote this kind of financing. Likewise, there are few intermediate institutions to improve the commercialization of technology. The existence of such intermediate institutions would likely increase either the creation of spin-offs or the innovation content of the incubating firms.

Conclusions

Cuernavaca and Ensenada are two medium-sized Mexican cities that have important capabilities in research and innovation development. Each has a champion research center that has played a leading role in innovation: the IIE (Electrical Research Institution) in Cuernavaca, and the CICESE in Ensenada. Both have promoted the creation of pioneering incubation centers in the region – CEMIT in Cuernavaca and IEBT in Ensenada – and they have created knowledge links with firms and other research centers: the IIE with a pool of resources for research in nearby Mexico City, and CICESE with some of California's research institutions in the US. Both TIPs gravitate around regions with a strong research infrastructure and together they form a northwest information technology area between Mexico City and Ensenada.

Cuernavaca and Ensenada's TBF industries are partially matched to their main research specialties. Cuernavaca's research specialty is in biotechnology, while local TBFs focus on software, biotechnology and energy. Ensenada specializes in telecommunications, while its largest TBFs focus on instruments, telecommunications and electronics. Both research centers are playing an innovative role, but CICESE is more actively involved in Ensenada than CEMIT is in Cuernavaca.

In sum, these research centers have played a pioneering role in innovation that can be emulated by others and facilitated and promoted by certain public incentives. Unfortunately, both TIPs' incubator firms went bankrupt after seven years in operation, apparently mainly because of a scarcity of venture capital for tenant firms. The challenge is to learn from the failures of the incubators instead of being influenced by the pessimistic attitude still prevailing among some actors in the regions. The failure came basically from an overestimation of the role of incubators and the lack of other institutional developments, mainly venture capital funds.

4.2.3 Large Cities: Monterrey and Guadalajara

Monterrey and Guadalajara are the capitals of the states of Nuevo León and

Jalisco, respectively. They are second and third in population in the country after Mexico City, with 3.5 million inhabitants in Guadalajara in 2000 (55 per cent of the state's population) and 2.9 million in Monterrey (76 per cent of the state population). Both industrial centers grew under the umbrella of the 'import substitution economic model' (1940–85), which established import taxes and other incentives to protect and promote industrial production for the internal market. As a result, their growth occurred originally in traditional industries such as foods and handicrafts in Guadalajara and breweries (1890) and steel (1903) in Monterrey. However, important changes have occurred since the opening of the Mexican economy in 1985, resulting in a more competitive environment and giving rise to a different kind of innovative firm.

The Monterrey innovation pole also includes the industrial corridor between Monterrey and Saltillo (state capital of Coahuila), along 70 kilometers of highway.

Regional technology infrastructure

The foundation in Monterrey of the ITESM (1943), a technical university sponsored by Monterrey's private industry, has encouraged the development of a local entrepreneurship culture.

In Guadalajara, the establishment of IBM in 1975 set the stage for the local computer industry. This industry grew to 90 000 employees and \$10.2 billion in exports in 2000. The telecom crisis of 2001–2002 caused exports to decrease to \$9.5 billion and employment to 78 000 by 2002.

In terms of technology areas, both poles focus on ecology and environment, machinery, manufacturing, foundry and software. In addition, Monterrey specializes in energy, chemistry and materials; and Guadalajara focuses on electronics, telecommunications, agribusiness and wood pulp and paper research.

Twelve research centers were selected for this study, seven in Monterrey and five in Guadalajara. Half of these are university research centers, one quarter are public CONACYT research centers, and the rest are private. In terms of size, half of the RCs are small and the other half are divided between medium and micro. So, neither city has any large RCs as the largest one is medium-sized, with 189 employees, the Public Research Centre in Chemistry (CIQA) located in Saltillo and dedicated to polymer research.

National researchers (NRS) in the RCs selected in the Guadalajara and Monterrey–Saltillo regions make up respectively 9.3 per cent of total NRS's researchers in the state of Jalisco, and 18 per cent of those in the state of Nuevo León. To have a sample large enough to be representative of this innovation pole, it would be necessary to also include research groups which are institutionally widespread. It must be noted, however, that the two states are relatively small and they account for only 3.8 per cent and 2.4 per cent of the country's NRS membership, respectively.

Both poles have more external knowledge links than internal ones. Monterrey has a ratio of three external to one internal link, and 56 per cent of its external links are international. Similarly, Guadalajara has an external to internal ratio of 2.3, and two thirds of its external links are international. Clearly, the RCs in both poles are very strongly connected with international RCs.

The links among research centers (RCs) show a chain of connections from Monterrey to Querétaro and finally to Mexico City, which acts as a link center, or the main source for other innovation poles. A second innovation chain, more specialized in biotechnology, originates in Mexico City and goes north through Querétaro to Monterrey (Figure 4.1).

Incubation mechanisms

Guadalajara and Monterrey have different incubation mechanisms: Guadalajara has a successful incubator and technology park while Monterrey has mostly indirect mechanisms for innovation.

Guadalajara's incubator, the University Centre for Technological Innovators (CUNITEC), was created in October 1992 as a joint effort of the University of Guadalajara (UdG) and the Nafin-CONACYT fund, which provided respectively 60 per cent and 40 per cent of its initial $550 000 assets.[42] One of CUNITEC's main objectives is to link university laboratories and research activities with industry. The incubator has 11 tenant firms out of a possible 20 at full capacity in a space of 4000 m^2. The incubator is located on the 8-hectare Belenes Industrial Park which houses 13 firms, six of which are TBFs. This industrial park was a government donation to UdG in the 1970s and the space is rented out by the university to raise funds and to encourage the transfer of university technology and services.[43] Firms in the park specialize in electronics, telecommunications, software, precision tools, mechanics, and light and clean manufacturing industries. From 1992 to 1997, CUNITEC hosted 23 tenants, eight of which graduated.[44]

The ITESM has expanded to 33 campuses nationwide. In 1985, it created an entrepreneurial program that has become a source of new ideas and spin-offs from the ITESM's incubation network created in 2002.[45] This network has given rise to 53 firms with 70 jobs, and 12 graduate firms distributed in 14 ITESM campuses located all around the country. Until 2001, Monterrey had no technology-based incubator.[46] At that time, the Physical Incubator of Businesses ('Incubadora Fisica de Empresas')[47] was established on the ITESM campus. This incubator hosted seven firms in 2003.[48]

TBFs surveyed

Fourty-six TBFs were selected for the study, both in incubation and not in incubation – 14 in Monterrey and 32 in Guadalajara. Forty-five per cent are micro-firms, 26 per cent are small, 13 per cent are medium, and only 15 per cent are large firms.

On average, Monterrey has more innovations per firm (4.2) than Guadalajara (2.7) (Table 4.4). However, Monterrey has a low coefficient of permanent environment for innovation which may be due to the fact that the city has no national labs. Guadalajara has a better indicator of constant innovation environment (Table 4.4). The Guadalajara industry with the most innovations is in ecology; in Monterrey it is in chemistry (five innovations per firm).

The analysis reveals that, in both TIPs, the local RCs and universities are important knowledge sources for TBFs. Monterrey's TBFs have more intensive relations with other firms outside the pole, suggesting that there is room for developing local technology firm suppliers. Guadalajara also has most of its knowledge flows with external firms, but it also has the support of local firms. Almost all firms are connected to national rather than international sources. This may be due to the fact that 85 per cent of their capital is nationally owned.

In summary, firms in both poles maintain links with local RCs that are connected with international knowledge flows.

A comparison of firm orientations and RC capabilities shows a pattern of technology mismatching between the two. Guadalajara's research specializations in information technologies (telecommunications, electronics and software) and biotechnology are well exploited by TBFs, but its large RC capabilities in machinery and equipment (4) and agriculture (5.5) remain untapped for innovations by firms in these sectors. On the other hand, the local firms' relatively stronger needs in ecology (3.5), telecommunications and energy (1.2) have only been addressed through limited number of RC projects (Table 4.4).

Monterrey's strong research center projects in ecology (2.7) and software (1.1), show no demand for their services from local firms in these sectors. However, the area's chemistry (3), biotechnology (2.6) and energy firms (0.7) have limited numbers of complementary RC projects (Table 4.4).

Overall, the interests of innovating firms are not well matched with the specialties of research centers in either pole. While Monterrey's RCs specialize mainly in ecology and software, its firms' technology areas are chemistry, biotechnology and energy. Guadalajara's firms focus on ecology, telecommunications and energy while its main research centers specialize in machinery and equipment, agriculture and electronics.

Both regions have a competitive advantage in terms of their diversified industrial demand and the high numbers of university graduates in the population. Monterrey has a vibrant entrepreneurial environment that could give rise to an innovative environment if venture capital funds materialize.

Both large industrial cities have a champion university to trigger innovation in their poles. The University of Guadalajara has been developing the Belen Industrial Park on its own, and has used its basic infrastructure to develop an incubator and a technology park (10 out of the 20 firms in the park were TBFs in 2002), the only technology park currently operating in Mexico.

Challenges for both innovation poles are, first, the lack of venture capital and mechanisms to promote its development and, second, the paucity of intermediate institutions to improve the commercialization of technology. One would expect the introduction of either of these to increase the creation of spin-offs or the innovation content of the incubating firms.

Conclusions

Guadalajara and Monterrey are the second largest metropolitan and traditional industrial cities after Mexico City, and they have had different experiences with the regional support of innovation firms.

Guadalajara has been developing information technology industries (electronics and software) since the pioneering settlement of IBM in 1975.[49] Local firms get their technology primarily from other national firms as well as from local RCs. Therefore the internationalization of Mexican RCs is positive in that it brings international knowledge to the attention of local technology firms.

Both poles have universities as champion institutions for innovation, but these universities have supported industry in different ways. Monterrey has been putting special emphasis on its entrepreneurial teaching programs. The University of Guadalajara has been developing the Belen Industrial Park, which has been used as base infrastructure for the development of an incubator supported by a special CONACYT program in the 1990s. Ten of its 20 tenant firms are technology-based, and, at least until 2005, it was the only operating technology park in Mexico. Its achievements still cannot be characterized as more than survival, due to the scarcity of risk capital for its TBFs.

A comparison of the technology specialization of RCs and TBFs shows that some capabilities are lacking or underutilized. In Guadalajara, more research is needed in the areas of telecommunications, biotechnology and energy, and local RCs could be more active promoting their technological expertise in machinery and software. In Monterrey, research and technology are well matched in information technologies and in material and equipment technologies, although the latter sector is generally weak. However, more research is needed to support TBFs in chemistry and biotechnology, and local firms could make better use of the existing research in ecology.

Monterrey has nearly twice as many innovations per firm as Guadalajara (4.1 and 2.3 respectively) (Table 4.3). Both poles have small innovativeness coefficients (though it is slightly greater in Guadalajara: 0.2), indicating weak institutional environments for innovation (Table 4.4).

The Monterrey pole lacks institutions promoting innovation, its only strong firms being in biotechnology and chemistry. The presence of risk capital and indirect mechanisms to promote innovation (incubators and science parks) could substantially improve its entrepreneurial climate.

4.2.4 Mexico City.

Mexico City area comprises the DF (Federal District), 37 out of 125 municipalities of the adjoining State of Mexico, and one municipality of the Hidalgo State, and is the national capital and largest city in Mexico (and possibly the world) (Figure 4.2). Historically, it has been the center of the country in many respects: political, economic, industrial, but also educational, scientific and technological. It has seen tremendous growth in both population (18.5 million in 2000, up from 2.9 million in 1941) and industry, the latter mainly oriented to consumer goods (the DF accounted for 29 per cent of the

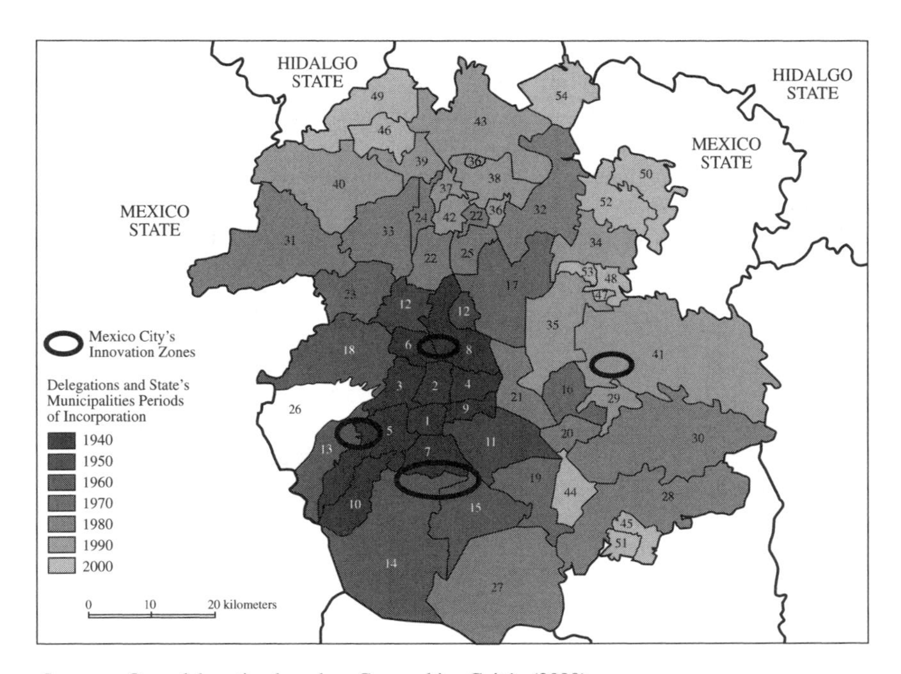

Source: Own elaboration based on Covarrubias-Gaitán (2000)

Figure 4.2 Mexico City's main innovation zones

industrial national product in 1998). Mexico City provides 18 per cent of the Mexican higher education and has 25 per cent of the nation's university graduates.

Regional technology infrastructure

Mexico City is host to four innovation zones: 1) The largest zone is in the south, around the National University, UNAM, up to Tlalpan where some National Health Institutes are located, and to the east with the Mexico City ITESM campus and the Metropolitan Xochimilco University (UAM-X); 2) North of the city, around the IPN, the oil national research institute (IMP) and the Science Research Institution (CINVESTAV); 3) To the east, an agricultural research zone based at Chapingo University and the Graduate Studies College (COLPOS), as well as the international research center for maize and wheat (CIMMYT); and 4) To the west, on the Toluca highway, the new zone of 'Santa Fe', based on consulting, educational research and high-tech services around the Iberoamerican University, the Santa Fe-ITESM campus, and the public center on economics and public policy (CIDE), among others (Figure 4.2).

Mexico City has half of the country's national research infrastructure in a variety of technological areas, such as pharmaceuticals, chemistry, instruments and energy. Ecology and services are also important research areas. The research carried out in this region supports national demands. The city has a major proportion of the country research capacity for research and development (52 per cent of NRS members were in the DF in 1999), an intrinsic result of its historical concentration of cultural, research and higher educational institutions.

Thirteen research centers have been selected for study. They are mostly medium sized, and among them, 11 pertain to universities, eight of which being part of UNAM, the largest university in the country, and three being at the IPN, also one of the biggest higher education and research institutions in the country. This number represents only a small fraction of the research centers located in the Metropolitan area as it covers 16 per cent of the NRS members in Mexico City. Nevertheless, these 13 centers have 606 NRS members (or 40 per cent) of the 1532 NRS members working in the 66 RCs included in this study (and presented in this chapter). Those 13 RCs are therefore much more productive than the 42 RCs examined in the other three cases plus the other eleven located throughout the rest of the country. With respect to knowledge fields, the DF accounts for 71 per cent of the nation's medical research and 64 per cent of national research in the social sciences and humanities. Research in the natural and physical sciences and engineering is slightly less centralized, with only 53 per cent carried out in the DF.

In contrast to RCs in other poles, Mexico City's RCs have more local/internal knowledge links, no doubt due to the great concentration of research in this pole. External links account for 40 per cent of the links and half are international.

Incubation mechanisms

Five incubators associated with universities promoted the creation of spin-offs during the 1990s, but only two were still in operation in 2002. The Centre for Technology Business Incubation (CIEBT) of the IPN, founded in 1995, had 17 projects in different incubation stages by 2002, and a more recent incubator founded by Mexico State University (UAEM) in 1998 had 10 firms in incubation. This incubator was located in Tecamac, a municipality in the State of Mexico. In 2001, it became a general-purpose incubator and, in early 2005, it may have no TBFs among its tenants. In early 2005, the CIEBT-IPN is the only one that continues to operate as a technology incubator in Mexico City.

Failed incubators include 1) the Incubation System of Scientific Firms (SIECYT) of the UNAM; 2) the Technological, Agricultural and Forestry Business Centre oriented to agricultural firms (Cetaf); 3) the 'Engineering Tower' multi-tenant building (Torre de Ingeniería) based on the concept of 'engineering associated firms'. Lessons from these failed incubators are described in the following paragraphs.

The 'Incubation System of Scientific Firms' (SIECYT) was founded in 1991 by the UNAM. It was one of the pioneering and better supported incubators during the 1990s (with Cuernavaca's CEMIT and Ensenada's IEBT incubators). It was located in a rented building in Tlalpan DF, south of the university campus. The UNAM incubator system had been developed as a university tool dependent on the Technology Innovation Centre (CIT) to promote and provide facilities for the development of technology-based firms, based on university technology packages.[50] It was closed by 1999, along with the CIT, because problems emerged in terms of operating UNAM's multiple knowledge links. After briefly trying to centralize these links, technology transfer became mainly the responsibility of each research center or school, only legal aspects being still managed by UNAM's central administration department for legal affairs.

The Technological, Agricultural and Forestry Business Centre (CETAF) was based at an agricultural university located in Chapingo, State of Mexico (UACH). It was supposed to start operations by 1994, but had a short life, and it failed probably due in part to the lack of involvement and hence support of the several other agricultural teaching and research institutions nearby (Colpos and the International Centre for Maize and Wheat, CIMMyT, as well as the central offices of the National Forest, Agriculture and Livestock Research Institute, INIFAP).

The Engineering Tower incubator ('Torre de Ingeniería') was developed by the Engineering Research Institute (II-UNAM) which has intensive technology transfer and research links, mainly with sponsorship of federal government agencies. The 9-floor, 7 million dollar tenant building was completed in 1999, with office space for 65 pre-selected firms. However, the student strike at UNAM from April 1999 to January 2000 thwarted the optimistic plans for this building, which would have included tapping into research services from the Engineering Institute and targeting professors and researchers who would eventually be part of those firms.

This high rate of incubator failure seems generally to be correlated with the absence of mechanisms to offset the inherent risk of innovation. For this reason in Mexico City's incubators may well be oriented toward general-purpose (rather than specialized) entrepreneurial support (as in the case of UAEM incubator) or they may be subsidized through university resources in case of technology incubators (as is the case at IPN incubator). Otherwise, without enough institutional sponsorship (like the failed UNAM incubator), they eventually fall into bankruptcy.

New incubators are on the way: Mexico City and Lago de Guadalupe ITESM-campuses in 2002; and the upcoming City Hall project of the High Tech Park 'Ferrerias', an effort to develop the northern high-tech zone of the city. The challenge, in these cases, will be to learn from the old experiences in order to make a better design, based in particular on sharing innovation risks among different actors and institutions.

Mexico City's current research institutions are mainly concentrated in the DF. For this reason, the local government is engaged in the construction of the aforementioned research park called 'Ferrerias' (located in the Azcapotzalco Delegation in the north of DF). The project is looking for high-tech firms in areas such as information technologies, new materials, telecommunications, electronics, robotics, scientific instruments, new energies and water technologies. It will also include incubators. The surrounding universities will provide technology and special services to support the park.[51]

In the early 1990s, incubators emerged through the promotional initiatives undertaken through the CONACYT Program, the main goal of which was to provide start-up support to TBFs. This left incubators with few or no financial resources to offset innovation tenants' risks, at a point when most tenants' projects were still only in their initial stages of development. This was a short-term view and many incubators were not sustainable and fell into bankruptcy. The result was that the number of firms graduating from incubation was quite low. For this reason, the technology incubators are changing their programmatic emphasis either to adopt a general entrepreneurship orientation (as the UAEM-Tecamac) or seek university or other financial resources (as is the case at IPN), or shut down (like the one at UNAM).

TBFs surveyed

A sample of 55 innovative firms has been selected in Mexico City. It represents only a small proportion of all such firms in the Metropolitan Area, but it is the second largest sample in the study after the one from the Querétaro–Bajío pole (see section 4.2.1).

Micro and small firms make up 63 per cent and large firms 27 per cent of the whole. These proportions constitute a different profile from that in the country's other poles, where large firms are the extreme minority. As in the other poles, however, the main technology knowledge sources for firms are other firms. In Mexico City, the relationship is primarily with internal firms (1.7 links per firm) and second, with firms external to the pole (1.4).

The technology profile is quite diversified. Eight of the 21 firms surveyed during the Monarca interview phase were in service (see section 4.2).

A correlation analysis explains 76 per cent of innovativeness (Table 4.3). On average there are 4.4 innovations per firm in Mexico City and the coefficient of 1.1 indicates a positive milieu of constant innovation. A few exceptional cases stand out, including 20 innovations in an electronic firm and eight innovations in a firm for transportation equipment and chemistry (Table 4.4).

The pole's research capacity is estimated by the number of research projects and adjusted with respect to national participation in each technology area (Table 4.4). Mexico City's technology capacity is highly concentrated in pharmaceuticals (12.8 times the national level), chemistry (11.7), instruments (10.8) and energy (9.1). Medium specialization exists in ecology (2.9), services (2.4) and machinery and equipment (2.3). Meanwhile, there is a relatively low technology specialization compared to the national average in other industries (1.5), including electronics and software (respectively 0.8 and 0.9), agriculture, and new materials (0.4 each).

On the TBF side, there is a diversity of minor specializations, including mostly pharmaceutical firms (1.5), as well as new materials (0.8) and instruments (0.5). These numbers suggest that, apart from pharmaceutical firms, links with local firms are very limited. This explains why the Mexico City research centers have a broad national orientation rather than focusing on Mexico City region only. These strong research capabilities (44.5) are poorly exploited by firms' demands locally (2.4) as well as nationally.

Conclusions

A more detailed zone-by-zone analysis would be necessary to depict the Mexico City innovation pole, which represents half of the research done in Mexico (52 per cent of NRS members are in the DF) and includes a broad diversity of scientific and technological fields. Yet the analysis shows that the results of the research done in this pole are disseminated to firms nationwide.

In this sense, the other Mexican innovation poles are partially sustained by Mexico City's research centers.

On a national level, the research specialization of Mexico City is highest in pharmaceuticals, chemistry, instrumentation and energy, and lowest in ecology, services and machinery. Interestingly, these specializations are not repeated in the pole's innovation firms, which show only a slight concentration in pharmaceuticals, and less in new materials and instruments.

Mexico City has great quantities of qualified personnel, as most of the nation's PhD programs are in the capital. However, this is changing as other regions are developing masters and doctoral programs, often with the support of universities in Mexico City, which are even reallocating research groups to areas such as Querétaro and Cuernavaca.[52]

The region's experience with incubation suggests that some members of the research community are interested in converting scientific results into technology. Yet innovation support mechanisms and financial funds are necessary to balance the risks involved in developing and transferring technologies and commercializing products and services. Until now, these mechanisms and funds have been lacking.

4.2.5 Conclusions

Mexico's innovation poles are distributed in two geographical 'chains' linking research centers and firms. Both innovation chains are linked to Mexico City. One angles to the northwest, passing through Guadalajara and up to the Ensenada innovation pole; the other heads north, crossing Querétaro and Monterrey, and includes Cuernavaca as well. The Bajío region is in the middle, potentially acting as a link between both chains. In contrast to this, the south and southeast of Mexico have very few innovation firms (Figure 4.1).

Mexico's innovation has taken place in five steps: (1) Up to 1958: centralized take-off in Mexico City, (2) 1959–76: growth of regional research infrastructure, (3) 1977–88: take-off of innovation firms, (4) 1989–94: growth of innovation firms, and (5) 1995–2000: decline of TBFs and of the research infrastructure.

The final decline of innovation seems to have been due in part to changes brought about by the opening up of the Mexican economy, but also in a large part to the absence of the financial resources needed to fuel the innovation process and to offset the risks associated with innovation investment. Most of the incubation and science parks in Mexico went bankrupt because of a lack of sufficient financial resources at a crucial stage of their development. In response, recent improvements in government policies have provided fiscal subsidies for up to 30 per cent of a firm's total investment in technology. Although the data is not yet available to assess the success of these policies, it seems probable that innovations in products and processes will increase when diversified sources of venture capital become available through a variety of channels.

Since most firms use other firms as sources of knowledge, future investigations should focus on the role of the latter firms, their relationship with RCs, and their ability to stimulate the former firms' development.

In general, mechanisms that could offset the risks of innovation and build a positive synergy stimulating development are still missing from innovation poles in Mexico.

Table 4.4 provides a comparative overview of the development paths and competitive factors of the selected knowledge regions in Mexico. These regions each have their own distinctive characteristics, but they share certain traits that existed prior to their development as technology poles. Except for Ensenada, they already had a skilled workforce and a good industrial and higher education infrastructure, of a size related to the size of the city, when their knowledge activities started to develop.

Incubators have been an important focal point for innovation in Mexico. They have been championed by different institutions in different regions, including the academic sector–main research centers (Cuernavaca and Ensenada) and universities (Bajío, Monterrey, Guadalajara and Mexico City) – and local government authorities (Querétaro's State Government). Incubator development has also often been backed up by partnerships among academia, industry and government, but even this collective involvement has offered too little support, and led to failures some years later (Ensenada, Querétaro, Cuernavaca).

Regions with a positive environment for innovation are supported by a good – and relatively large – network of research centers (Querétaro–Bajío, Mexico City). Some regions have proactive research centers (Cuernavaca, Ensenada), which promote technology transfer and spin-offs by their researchers, while innovation in big and industrialized cities is usually triggered by a university champion institution (for example, ITESM in Monterrey, and University of Guadalajara in Guadalajara).

Each innovation region has strengths in specific fields of technology that do not always correspond to their innovation firms' areas of specialization. Cuernavaca is strong in biotechnology research and Ensenada in telecommunications research, yet both have only a small number of TBFs involved in these fields. Instead, Ensenada's high-tech industries focus mainly on instruments, telecommunications and electronics, and Cuernavaca's industries specialize in software, biotechnology and energy. Monterrey's RCs specialize in ecology and software, while the firm tech areas are in chemistry, biotechnology and energy. Guadalajara's firms focus on ecology, telecommunications and energy, while its research centers' main technology fields are machinery and equipment, agriculture and electronics. One possible explanation for this mismatching of abilities is that firms demand relatively little technology and research from research centers, and the latter are therefore oriented to national rather than regional needs. This is true for Mexico City, which has a diversified, high intensity research infrastructure concentrated in pharmaceuticals, chemistry, instruments and energy, as well as ecology and services. Mexico City's strong research infrastructure functions as a 'research node', oriented to national research demands and demands from other research centers. Another possible explanation is that the firms in the innovation poles have followed the global technology trend, targeting biotechnology and information technologies rather than local specialties.

Mexico lacks public policy measures needed to increase the supply of venture capital and incentives to encourage risk capital investments into firms' innovation processes. The absence of mechanisms that would offset the inherent risk of innovation has led to incubator failure. Incubation activities in Mexico are currently initiated and supported in one of two ways: by established firms, research centers and/or government initiatives; or by universities, which have entrepreneurship courses that aim to increase incubation activities. The ITESM is an example of the latter, with a plan for 25 incubators which could find tenants from firms that emerge from entrepreneurial teaching activities at campuses; and the UNITEC, another private university based in Mexico City, which sponsors virtual business on the internet by doing annual calls for tenant firms (they had the 4th call in 2004) as well as offering courses and specialized consulting services in entrepreneurship. Entrepreneurial activities and courses at the universities are widespread, but only in a few cases do they lead to innovation in the production sector or to the creation of new firms. CONACYT's graduate scholarship program has been an important step in the effort to develop research talent, but it has suffered from the lack of complementary programs to attract and retain young graduates with master's degrees and PhDs, largely because research positions are few and far between, especially in the private sector.

The regions' sustainable development depends on the collaboration between the three main agents: (1) one or all three levels of government (national, state and local); (2) universities; and (3) the private sector. The commercialization of technology requires strong channels for knowledge flow, and also some support policies related to SMEs, including nurturing and supporting spin-offs and encouraging labor mobility, in particular of researchers and qualified technology personnel.

Each of these three agents can initiate the development of an innovation pole. The government (local, state and/or national) has been the proactive institution in the case of Querétaro (similar to US case of Evanston and

Quebec, Saskatoon and Calgary cases in Canada). A local university has played that role in Monterrey, Guadalajara (as is the case in the New York capital region). This role has been taken on by a research centre in Cuernavaca and Ensenada. In Ottawa, it is the private sector, taking advantage of government R&D, that initiated local high-tech development. Regardless of which institution is proactive, the important issue remains to facilitate knowledge links and encourage cooperation or partnerships between all three types of institutions: governments, universities and private firms.

Mexico must face the challenge and promote regional innovative environments. In order to do so, it must (1) tackle innovation gaps and find ways to link regional research with TBF creation so as to produce a leapfrog innovation effect; (2) consider technology policies that discourage the borrowing and recycling of ideas and encourage the creation of new technologies; (3) develop an institutional framework for internationalizing innovation links;[53] and (4) create a policy that promotes the development of venture capital funds.

NOTES

1. The population was 97.5 million in 2000 (INEGI, 2001). By 2003, it was 102.7 million inhabitants (OECD 2004c).
2. 'Tertiary-education level' means 'post secondary' studies, including universities, technical colleges, continuing education, and professional development.
3. The OECD figures for Mexico are 39 736 R&D personnel in 1999, and 43 455 by 2001; this marks an increase to 1.1 R&D personnel per thousand employees (OECD 2004c).
4. 1997 data.
5. Economist Intelligence Unit (2005).
6. International Trade Statistics, 2001, (www.wto.org)
7. A maquiladora or in-bond firm is a manufacturing plant making products or parts that are imported temporarily on behalf of a third industry or service firm. It is entitled by the Mexican government to singular 'customs treatment, allowing duty free temporary import of machinery, equipment, parts and materials, and office equipment subject only to posting a bond guaranteeing that such goods will not remain in Mexico permanently. Ordinarily, all of a maquiladora's products are exported, either directly, or indirectly, through sale to another maquiladora or exporter. The type of production may be the simple assembly of temporarily imported parts; the manufacture from start to finish of a product using materials from various countries, including Mexico; or any conceivable combination of the various phases involved in manufacturing, or even non-industrial operations, such as data-processing, packaging, and sorting coupons'. (See 'Decree for Development and Operation of the Maquiladora Industry', Diario Oficial, 22 December, 1989, http://www.bancomext-mtl.com/invest/vox128.htm.
8. *Hacienda* is the common Mexican word for a large landed property with a production unit. *Haciendas* were common from the colonial period until the expropriation of lands during the Mexican Revolution (1650–1940). *Peones*, who worked on *haciendas* doing agricultural, mining, or domestic activities, were paid with lodging, food or clothing, things that were valued on a payment list (*lista de raya*).
9. The NRS gathers researchers who receive a positive evaluation based on their scientific publications and their contribution in terms of Master's and PhD students' theses.

10. This typology is adapted from two sources: (1) the functional role is stated in the pole innovation model (Figure 1.1); and (2) some categories are proposed by Casalet (2000): a) organizations providing incentives, b) highly specialized R&D centers, c) human resources development, and d) institutions providing information and reducing uncertainty regarding standard, quality and training.
11. The ENESTYC (survey on employment, salaries, technologies and training in the manufacturing sector) is a general statistical source that looks for technology changes by industry. Domestic sources of technology are estimated at 7 per cent while spending on foreign technology represents 93 per cent (Unger and Oloriz, 2000, p. 88).
12. The number of research centers in Mexico can be estimated at about 350. Some of these research centers are located in universities, either as established institutions or working groups, and these include 100 technological research centers. Our sample of 66 thus provides a good picture of the whole.
13. Martínez Rizo (2000) Some of the problems affecting Mexican universities are:
    - Directors have no management training.
    - The rate of student graduation is low.
    - Links between universities and industry are weak, basically related to research.
    - It is increasingly difficult to control the quality of education because of the growing number of universities, the increasing share of private education, and the diversification of degrees.
    - Funding shortages generate conflicts of interest between pursuing academic activities and selling services.
    - Facing global competition from international universities (Gacel-Ávila, 2001).
14. The fiscal incentive grew from $53 million to $91 million from 2000 to 2004, affecting 300 to 562 firms. It may be noted that in the US and Canada the incentive came as a deduction from taxable earnings when a company files its annual income report, unlike in Mexico where it is a competitive program, so it is not available to all firms doing R&D.
15. The CGCyT and other related laws were approved in 2002.
16. The 'sectors funds' operate in partnership with the federal government agencies, collecting $127 million in 2003.
17. The Conacyt programs are: 1) The Technology Modernization Program, PMT, which supports SME's technological improvements; 2) The Program for the Support of Research and Development Projects (PAIDEC), which encourages firms to tap into knowledge-resources available at universities; 3) Outreach program PROVINC, which supports programs undertaken jointly with the academic sector and creates technology services and management units (UGST), as well as productive sector council boards (CASP), that work to increase the interest and capability of universities to service demands from the productive sector (www.conacyt.mx/dat/avance); 4) The Research and Development Fund for Technology Modernization (FIDETEC), which invests in innovation and technological development projects when they are in the pre-commercial stage.
18. Source: Conacyt (2004).
19. The regional programs objective is to promote research on a regional level; to encourage the production system to participate in the financing and use of research results; and to build decentralized R&D capabilities and regional cooperation in science and technology (Ponce Ramirez, Luis, 1999).
20. Conacyt's regional budget is changing as nowadays are merely defined for administrative purposes, as well as the configuration of certain regions. Three states – Morelos, the State of Mexico and the DF (Federal District) – were not considered to be regional research systems, but they provide 65 per cent of the national researchers and contribute 35 per cent to the GDP. By 2005, the Conacyt funding activity was mainly carried out by each state with the so called 'mixed' funds (http://www.conacyt.mx/dadrys/directorio/index.html).
21. The national researchers in the other states housing the selected four cases made up different proportions of NRS: Morelos (6 per cent of the NRS), Guanajuato and Querétaro (5 per cent of NRS) and Nuevo León and Coahuila (4 per cent of NRS). These values are calculated for the 2003 year based on Conacyt data (www.siicyt.gob.mx).
22. The number of NRS members of each of the regional systems can be correlated with its

Regional Gross Domestic Product (Table 4.1), as 92 per cent and 90 per cent are the statistical explanations obtained for 1996 and 1999, respectively. The relationship between regions is that a unit increment of GDP national share nearly doubles the national researcher's share (for 1996 and 1999). The correlations are the following:

NRS(%) = - 8.83 + 1.97 (GDP%), year 1996; R^2 = 92%
NRS(%) = - 8.03 + 1.88 (GDP%), year 1999; R^2 = 90%.
Where, NRS (%) and GDP (%) are national shares.

23. Other innovation poles in which TBF and research centers are located are the cities of Toluca, Chihuahua, Puebla, Veracruz, San Cristobal (Chiapas) and the region of Mar de Cortes.
24. Considering both the population size definition described in section 4.1.1 and the proposed urban classification by Rodriguez, J. y and M. Villa (1997), the following urban limits can be integrated:
    Metropolitan cities, more than 4 million inhabitants;
    Big cities, less than 4 million and more than one million;
    Medium cities, from 100 000 up to 1 million.
    Small cities, with less than 100 000 and more than 15 000,
    Semi-rural areas, less than 15 000 and more than 2500 inhabitants.
    Small rural towns, with less than 2500 inhabitants.
25. This crisis began in December 1994, apparently due to a government cash shortage to cover international obligations. The Mexican peso devaluation reached 5.15 pesos per dollar from 3.5 pesos 15 days before (16 December 1994 to 30 December 1994). The devaluation process continued during 1995 up to 7.7 pesos per dollar. That is 120 per cent devaluation for the whole year, causing the economy to shrink to (−6.2 per cent) in 1995. In the ten years since this crisis Mexico's annual GDP growth has averaged a mere 2.6 per cent, and only 1.1 per cent in per-capita terms (*The Economist*, 2004).
26. The total number of interviews conducted at the TBFs was 262, including 116 during the Indico1 period, 43 during the Indico2 period, and 103 during the Monarca period. Some TBFs were interviewed more than once, mainly during the second period (Indico2), as the goal was to understand the impact on them of the 1995 economic crisis. Overall, 350 interviews were conducted, in such a way that 184 TBFs were interviewed once, 68 twice and 10 three times.
27. The large firms each with a RC are CONDUMEX (auto parts), TREMEC (electrical cable), MABE (home supply) and HOUSE HOLD (plastics: mold design and injection processes).
28. The size of the research centers is defined in relation to the number of total employees. The scale chosen is the same as in the firms (see note 33).
29. The graduated firms are: INSIDE, operating in link services, Carlos Pirsch, solving environmental problems; and de Salti, product development for MABE Corporation.
30. CIO (Centre for Research in Optics), CIATEC, is a technology center oriented to the shoe & leather industry.
31. The stakeholders of the Incubators are: Conacyt, NAFIN, State Government, each with $300 000 dollars, and CIATEC, CIO and the University of Guanajuato, each with $30 000. The municipality of Leon gave 10 000 square meters of land.
32. The four firms are: 1) dealing with construction material, 2) rescue equipment, 3) software development for the shoe industry, and 4) software development for numerical control equipment.
33. The firm scale used in Mexico for the firms' sizes are: a) 'Micro' (meaning very small) having less than 15 employees, b) Small firm with less than 100 employees, c) Medium, less than 500 employees and d) Big, more than 500.
34. The special cases of single firms, with 15 and 10 innovations in new material and instruments, are not significant, even though they can probably show a kind of industry based on science. This typology comes from Pavitt (1984) who classified industries from the point of view of innovations as: 1) dominated by the suppliers, 2) scale-intensive, 3) specialized supplier, and 4) based on science.
35. Research centers have generally diversified research activities, making it more difficult to be classified by industry or technology area. Therefore, instead, their research projects have

been chosen which have been classified by industry. For TBF either the innovations or the TBF can be classified by industry, and then both are good proxies for innovation results. But from the point of view of the size of the sample, the three total TBF enquiries can be used (see section 4.2), but for innovations there is better information for Monarca, which is the last enquiry.

36. The specialization coefficient is calculated in Table 4.4.
37. Another incubator experience was that of Mexicali, the SETAI (Incubating System of Technological and Industrial Businesses) founded in 1993 by the State University (UABC). This incubator has been in stand-by operation for a couple of years.
38. The fund is called Morelos Technology Park Trust, sponsored by the State of Morelos, a local industry association representation, three national universities (UNAM, IPN and UAEM), the National Science & Technology Agency (Conacyt), a financial organization (Nafinsa) and a research center (IIE).
39. August 1990 is the foundation date of IEBT at Ensenada.
40. Some of the graduated firms from the incubator in operation by 1998 are: Biopesca founded in 1991, Ambientec (1992), Acuacultura Oceánica (1993), Acuícola San Quintín (1995).
41. Also in Bajío's agriculture sector the largest number of innovation per firm is found (6.7) (see section 4.2.1).
42. The university funding source corresponds to the installation's value, while the Conacyt is exhibited in cash during 1993 and 1994. This has been calculated with an exchange rate of $3155 pesos per dollar.
43. The University of Guadalajara, the second largest Mexican University, has a larger Technopole project that includes 23 hectares.
44. CUNITEC has changed its name to IEBT-UdeG and is hosting 12 firms and organizations, and the Belenes Industrial Park has 10 technology-based firms out of 20 (updated by Dr Juan Villalvazo, university officer in charge of the administration, February 2002).
45. By 2004 there were 500 firms in incubation at 25 ITESM campuses (http://cde.itesm.mx/red_directorio.php).
46. There is a commercial incubator run by the City Council with the sponsorship of the ITESM.
47. Incubadora Física de Empresas, stands for a kind of Bricks and Mortar Incubator.
48. The incubator has been considered a non-feasible project for a long time, according to a study which recommended that an incubator should not be developed in the ITESM-Monterrey (Musalem, 1989). However, by 2002, ITESM has launched a network of incubation activities spread nationally over its 33 campuses (see Table 5.4).
49. IBM was inaugurated in 1975 and began producing PCs in 1986. This year Wang also produced computers. In 1989 Compubur (associated with Unisys) was created. Also some Mexican firms such as Logix, Electrón, Mexitel and Tandem along with some suppliers of computing equipment firms such as Adetec, Cherokee, Pantera, Molex, Encitel, and NM were established (Partida Romo, 1999).
50. Siecyt has been approved by the Technical Council of Scientific Research (Coordinación de la Investigación Científica) of UNAM on 13 September, 1991.
51. The North Mexico City Institutions are IPN, UAM-Azcapotzalco, Technical training public schools (Cebetis, Cecatis) and Health Hospitals.
52. The Juriquilla campus in Querétaro has been a place to move some UNAM and CINVESTAV research groups in applied physics, neurology, geosciences and management which are an extension of the research institutions.
53. Internationalizing the links for innovation, the Mexican government supported the opening of the incubator 'Technology Business Accelerator of Mexico', in San Jose, CA, in January 2005. This new incubator includes 25 Mexican companies in the fields of software, pigments, mechatronics and hardware, and is trying to take advantage of access to the knowledge and networks available in Silicon Valley <www.techba.com>.

5. Comparative analysis of the selected North American knowledge regions: lessons and conclusions

INTRODUCTION

The fourteen North American technology innovation poles (TIPs) described in Chapters 2, 3 and 4 form a very heterogeneous sample. The diversity in their development profiles may be attributed to several factors: their respective national culture and regional norms; their socio-economic environment and system of innovation; the local conditions and resources present at the outset of development; respective economic transformation agendas; local organizations, local leaders and innovation institutions. It may also be due to the timing of development with respect to economic and business cycles or even to the development process itself. The objective of this chapter is to learn from these differences by analyzing the prevailing conditions, key regional characteristics, mechanisms employed, and decisions made during the course of growing these TIPs. These lessons may provide useful insights and compelling evidence as to why some TIPs are more fortunate than others in developing, attracting and retaining technology-oriented firms.

The chapter starts with a comparison of the NAFTA countries' national systems of innovation. As outlined in Chapters 2, 3 and 4 and summarized below, there are significant differences between the three countries in terms of socio-economic conditions in which science and technology, talent and capital issues are addressed and how these differences affect their regional development. This is followed by a comparison of the twelve TIPs examined in this book and by an analysis of the characteristics and effectiveness of some of the incubation mechanisms used in each region. The chapter ends with perspectives on innovation and TIPs in each country and in NAFTA.

5.1 NAFTA COUNTRIES' INNOVATION CAPABILITIES AND TECHNOLOGICAL PATHS

When one looks at the economic level of the three NAFTA countries, their

Table 5.1 The socio-economic and NIS context of NAFTA countries

Countries	Population 2002 numbers (millions)	Population 2002 (%) NAFTA share	GDP 2002 (US Billion PPP$)	GDP per capita 2002 (US PPP$)	GDP per capita 2002 (Mex = 1)
Canada	31.41	7.5	951.93	30,306	3.29
Mexico	100.44	24.0	925.5	9,214	1.00
US	287.46	68.5	10383.1	36,120	3.92
NAFTA	419.31				

Countries	GERD 2002	GERD 2002	% financed by		% performed by		
	(Millions $ PPS)	(Mex = 1)	Industry (%)	Government (%)	Industry (%)	Higher education (%)	Government (%)
Canada	18 447	5.2	44.3	34.0	53.7	34.9	11.2
Mexico	3565	1.0	29.8	59.1	30.3	30.4	39.1
US	284 584	79.8	63.1	31.2	68.9	16.8	9.0

Country	Number of researchers full time equivalent 2002	Number of researchers full time equivalent per thousand employment 1999	GERD 2002 (% GDP)	GERD per capita population 2001 (PPP$)
Canada	107 300	6.6	1.91	593.5
Mexico	21 879	0.6	0.39*	36.0
US	1 261 227	8.6	2.67	964.7

Note: * Year 2001

Source: OECD 2004b

institutional framework and policies, their scientific and innovation activities, the US generally places first, Canada second, and Mexico a distant third. The differences this ranking indicates are reflected at the regional and local levels and affect the evolution of knowledge regions. It is therefore not surprising that each case analyzed in Chapters 2, 3 and 4 shows a unique development path leading to differing outcomes.

In 2002, NAFTA had 420 million inhabitants, 68.5 per cent in the United States, 24 per cent in Mexico and 7.5 per cent in Canada (Table 5.1). GDP measured in PPP dollars[1] was $10 383 billion in the US, $952 billion in Canada and $926 billion in Mexico, which in per capita terms was $36 120 in the US, $30 305 in Canada, and $9219 in Mexico; US being 3.9 times and Canada 3.4 times to that of Mexico.

The US and Canada have relatively well educated populations and technologically-minded cultures, as reflected by the high proportion of the working-age population that has a post-secondary education and the number of researchers and R&D expenditures per capita. Mexico is still far behind its NAFTA partners on these dimensions, even though it is actively developing its post-secondary education sector. In 2002, only 8 per cent of Mexico's population between the ages of 15 and 64 had tertiary-level education,[2] compared with 37 per cent in Canada and 32 per cent in the US (OECD, 2004a, p. 22). In terms of advanced degrees, the proportion of the population at the typical age of graduation who received a PhD degree in science or engineering was 0.5 per cent in the US, 0.3 per cent in Canada, and 0.004 per cent in Mexico in 2000 (OECD 2003, p. 51; Conacyt and INEGI data). Mexico also lags far behind the US and Canada in terms of the number of science and technology researchers: in 2002, there were 1.26 million researchers in the US, far more than in Canada (107 000) and Mexico (22 000). In relative terms, in 1999, the US had 8.6 researchers per 1000 jobs in the labor force, compared with 6.6 in Canada and 0.6 in Mexico (Table 5.1).

The US is generally considered to have one of the best established, most complex, and extensive innovation system in the world. In 2002, its gross expenditures on R&D accounted for 43.85 per cent of all of the OECD's (OECD, 2004b). Canada's research infrastructure is also well developed, while Mexico's is still quite weak. In 2002, the US's gross domestic expenditure on R&D (GERD) was estimated at $284.6 billion, Canada's at $18.5 billion and Mexico's (in 2001) at $3.6 billion (Table 5.1). In absolute terms, Mexico's R&D activity is therefore about one fifth of Canada's and one eightieth of the US's.[3] As a share of GDP, GERD stood at 2.74 per cent in the US, 1.92 per cent in Canada, and 0.39 per cent in Mexico in 2001. Mexico's target for 2006 is to attain, 36 years later, Canada's 1970 R&D level of 1.0 per cent of GDP. Gross per capita expenditures on R&D in 2001 were PPP$965 in the US, PPP$594 in Canada, and PPP$36 in Mexico (Table 5.1).

In the US, R&D is clearly driven by industry and the non-profit private sector, which are also involved in a number of research partnerships with higher education. The situation is similar in Canada, where a major share of the country's R&D activities is performed directly by industry, and university–industry research partnerships are very well developed. As in the US, the industry share of R&D has been growing steadily in Canada over the past decades and is expected to continue this trend for the coming years; the country's technology policy has been to shift responsibility for applied research and its commercialization to the private sector. In Mexico, the government is still a dominant player in R&D, either directly or through higher education. In this context, industry performs a growing – but still small – share of the country's R&D activities. In quantitative terms, in 2002, industry financed 63.1 per cent and performed 68.9 per cent of the national R&D in the US, while the numbers were respectively 44.3 per cent and 53.7 per cent in Canada and 29.8 per cent and 30.3 per cent in Mexico (Table 5.1). Mexico's industrial R&D is clearly weak. These differences are due less to chance than to different levels of education and industrial development, as well as the different stage at which each country's national system of innovation finds itself.

Most public research programs involve partnerships with the private sector, leveraging public funds with industry funding. Governments are also offering generous research tax credits and capital gains tax treatment to encourage private sector research. In 2001, the rate of tax subsidy for R&D was 7 per cent for firms in the US, 4 per cent in Mexico and 18 per cent for large firms and 32 per cent for small firms in Canada (OECD, 2003b, p. 43; also see note 14, chapter 4). As expected, in all three nations, there are regional differences within each country in funding designed to promote specific developmental objectives. Besides this, the availability of venture capital is uneven, concentrated in a small number of more visible regions. Venture capital is generally considered to be a key element in the creation and growth of technology-based firms (TBFs) and the commercialization of new technologies. In 1998–2001, VC investment as a percentage of GNP stood at 0.49 per cent in the US (with around one third going to advanced technology firms), and at 0.25 per cent of GNP in Canada (with around 80 per cent going to advanced technology firms; OECD, 2004b, p. 18). In contrast, venture capital investment in Mexico is still very low.

After World War II, US R&D policy gradually shifted its emphasis toward civilian technology transfer with tangible programs starting to appear in the 1960s. With the end of the Cold War in the late 1980s to early 1990s, this focus has shifted towards cooperative technology development, in which new networking-oriented activities are based on complex and interactive partnerships that affect the development and diffusion of technology. R&D

policies have promoted newer patterns of industrial collaboration and commercialization, including industry consortia, university–industry linkages, and public–private partnerships. Science parks and technology incubators have also been important mechanisms for innovation, promoting networking and interactions between science, technology and industry. As outlined in Chapter 2, a key element of the US innovation system is the important role it has given to small and medium sized enterprises (SMEs) in the commercialization of technologies developed during the intensive defense R&D periods and in the development of appropriate technology transfer mechanisms. Indeed, SMEs have been major players in the innovation process, as producers, adapters and 'commercializers' of new technologies. As a result, small and emerging technology-based start-ups have played a significant role in the development and diffusion of microelectronics, computer hardware and software, biotechnology, and robotics in the US during the past five decades. US innovation programs include initiatives that encourage multifaceted government–university–industry research partnerships, initiatives to facilitate access to technology by SMEs, and help establish mechanisms to provide incubation support to technology-based businesses. Many of these technology-based economic development (TBED) initiatives have been funded through federal grants and a combination of state and local matching funds often implemented with active private sector involvement. These projects have had mixed results, affected as they are by the larger economic trends, federal, state, and local politics and the success and failure of participating TBFs. In spite of the ups and downs in political support and ensuing funding levels, often the state and local commitment with creative local and regional coalition-building have contributed to the continuing popularity of TBED programs in various US regions (Chapter 2).

In Canada, the federal government financed over half of all the country's research activities until the late 1960s, R&D activities being performed in almost equal proportions by government laboratories, industry, and the universities. Since that time, policies and programs have been developed that seek to: increase the level of R&D activities in the country, increase the role of the private sector as a source of funding and a performer of applied research, develop stronger university–industry research linkages to enhance the relevance of university research, increase the commercialization of research, enhance technology transfers to SMEs, by facilitating their access to technology for development and commercialization of new or improved products (Chapter 3). To that end, a number of industry partnership programs have been created, at both the federal and provincial levels, leveraging private research with public funds, facilitating joint university–industry research, and encouraging collaborative work between private firms and public laboratories. In addition, an extensive infrastructure has been developed to facilitate SMEs' access to technology which includes research funding programs, centers of excellence that facilitate access to university research, industry partnership facilities in NRC research laboratories, and a proactive industrial research assistance program that provides research funding and advice by an extensive network of industry technology advisors. These programs – combined with supportive tax credits for R&D activities and in collaboration with local economic development organizations and regional industry–university–government consortia, actively promote networking and partnership activities – help to create environments that are supportive of innovation. Technology investment and commercialization would be further enhanced in Canada if a more favorable tax treatment of capital gains was introduced to reward investments in emerging and growing advanced technology firms. Overall, the science and technology system in Canada is not very different from that in the US, with its current emphasis on industry–university–government research partnerships to promote the commercialization of technology, and the inclusion of SMEs as key elements of the innovation system. Early attempts, in the late 1970s and early 1980s, at building formal incubators and incubation spaces had met with only limited success. A wave of formal incubation mechanisms have recently started once again to spark interest in universities and research laboratories, where they build on the culture of partnership, cooperation and networking that has been developed in the recent past.

Mexico's long history of creation and loss of technology capability has followed two lines of development, one based on indigenous specialties (medicine, astronomy, construction, architecture) and the other on technologies imported since the Spanish Conquest in 1525 (Chapter 4). Both lines have evolved under economic and social conditions that led to the use of labor-intensive rather than productivity-enhancing technologies. Today, this approach is illustrated by the 'maquiladora' industries, which are based mainly on the supply of cheap labor and therefore keep local demand for innovation in the form of new technologies and applied knowledge low. The key milestones in the emergence of a system of innovation in Mexico include: the creation of the Mexican Petroleum Institute in 1965, which was followed by the creation of research institutes in nuclear science and electricity in 1972 and 1975, respectively, and the establishment of CONACYT as a coordination agency for science and technology in 1970. Before CONACYT, the Mexican scientific activity and its coordination was concentrated in Mexico City. With Conacyt, a regional research infrastructure has been created with research centers in several regions and the active development of higher education programs in science and engineering. A number of advanced technology firms emerged in the late 1970s and 1980s and did well initially. In the mid-1990s, however, the level of scientific research and the rate of creation and growth of

TBFs started to decline. This happened in spite of – or perhaps because of – the new economic model implemented with NAFTA. In order to reverse this decline, one possible alternative may be to blend modern technologies and competencies imported from abroad, with local capabilities. This would involve the development of regional systems of innovation around selected technology niches based on endogenous knowledge and technologies that would be open to external research relationships.

Overall, the qualitative differences between the US and Canadian innovation systems are not very significant, as both countries have somewhat similar structures and overlapping TBED priorities, including in particular government–industry–higher education partnerships and technology transfer to SMEs. Canada's objective is to increase its R&D efforts by creating conditions that will encourage more industrial R&D, thus increasing its similarity with the US. Mexico is in a different situation, with relatively low R&D activity, few scientists, limited venture capital, and a less well educated population than the US and Canada. It will take many years for Mexico to reach levels comparable to those in Canada and the US, whether in terms of science and technology infrastructures or the supply of researchers and R&D spending. As a result, Mexico may need to find its own way ahead instead of importing a 'foreign' science and technology system. It should probably focus its efforts on a limited number of regions and sectors, building on local innovation capabilities while at the same time increasing its scientific human resources and research activities, developing its research base, and creating policies that would lead to increased industrial research in large and small firms. Mexico's best chance to accelerate its technological development is tied to the focused use of its limited scientific resources. Its technology transfer activities must therefore be tailored to the small size of its scientific and technological infrastructure and its local innovation capabilities.

5.2 BUILDING UP OF THE KNOWLEDGE REGIONS AS TIPS

As outlined in Chapter 1 and illustrated by the case analyses of Chapters 2, 3 and 4, TIP characteristics change over time as they take off, grow, mature, and eventually stagnate or decline. The 14 NAFTA regions covered by our analysis are not currently at the same stage of development, even though the key decisions that led to the development of most of them were taken around the same time period, in the late 1970s and early 1980s. Their evolution has been affected both by external and internal factors, including international business and technology cycles, their national socio-economic and regulatory environments, and local characteristics and decisions.

Some of the characteristics of the 14 knowledge poles described in Chapters 2, 3 and 4 are summarized in Tables 5.2, 5.3 and 5.4, as they evolved from their traditional socio-economic base until current times: pre-conditions, take off, current knowledge-based clusters, formal incubation infrastructure, overall assessment of current conditions in terms of the area's performance as a competitive and sustainable knowledge pole, and current challenges. The comparative analysis that follows reflects mainly current conditions. It is based on the analytical framework presented in Chapter 1 and depicted in Figure 1.1, which focuses on: key regional actors, the regional context, resources, institutions, and the outcomes in terms of innovations, nascent and established TBF clusters and high-tech jobs

Based on the information presented in Chapters 2, 3 and 4, Tables 5.5, 5.6 and 5.7 summarize the presence and/or impact, on innovation and existing and emerging TBFs, of the characteristics considered in our analytical framework. The tables show these characteristics' comparative presence/impact in a given region (strong, medium or negligible), as it compares to the most successful regions in the country. These intra-country estimates by the authors are approximate measures of the current level of the regions' development and success within their country. In each country, a region with all characteristics judged to be 'equal to the best in the country' could have a maximum score of 54 (that is, the same number of TBFs as the largest TIP in that country, same ease of access to venture capital as the TIP which received the most in that country, same quality of life, and so on). Hence, all of the 18 items multiplied by 3, which is the highest score, gives 54. When even the 'best in the country' was judged to be deficient (such as financial resources in Mexican regions), a lower score was given.

Within each country, a clear ranking of each region in terms of advanced technology development and sustainability emerges from the observation of Tables 5.5, 5.6 and 5.7:

- In the US, the NY Capital Region and Chicago Metro Area come first as strong TIPs, followed by Madison Metro and New River Valley, in that order.
- In Canada, Calgary and Ottawa come first, followed by Saskatoon, and Quebec City.
- In Mexico, Guadalajara and Monterrey come first, followed by Mexico City and Querétaro–Bajío in that order, with Ensenada and Cuernavaca showing up far behind.

Straight inter-country comparisons cannot be made because of the significant differences between countries in terms of innovation systems, level of economic and industrial development, and cultural aspects. Useful

Table 5.2 Comparative overview of the development path of selected US knowledge regions

Development characteristics	New York Capital Region	Chicago Primary Metropolitan Area	New River Valley Region, Virginia	Madison Metropolitan Area, Wisconsin
Pre-conditions ● Area's importance ● Skilled manpower ● Higher education ● Private industry ● Other	Seat of New York state govt. History of industrialization with skilled, educated and disciplined workforce. Presence of RPI, SUNY-Albany & several technical colleges. Private research labs including GE R&D Lab, Sterling Winthrop Research Inst. Moderate to high quality of life amenities.	Presence of Chicago, a stable well diversified economic base and int'l business & communication hub. Skilled and educated labor force. Presence of research universities – Chicago, Northwestern, Illinois & others. Numerous large firms headquarters and R&D labs. High quality of life with amenities.	Area's burgeoning population due to favorable climate and central location; the presence of a major research university with R&D-intensive environment prompted to plan for economic diversification. This rural area was lacking any major private tech oriented businesses and univ. tech transfer offered numerous opportunities.	The UW's history of involvement with patents and tech transfer program such as WARF; shrinking public financial support for the university, prompted the establishment of innovation infrastructure. It was hoped to aid commercialization of univ. research for financial gains.
Take off ● Restructuring need ● Infrastructure dev. ● Lead actors ● Other milestones	Area experienced downturn in smokestack industries during 1970s/80s triggering layoffs; several GE spin-offs by former employees. RPI Univ. with visionary leadership takes lead in setting up innovation infrastructure – incubator (1980), science park (1982), & several inter-disciplinary research centers.	In early 1980s, region's stagnant economy and higher taxes particularly in Evanston warranted the need to leverage university technology to help diversify area's service economy, create jobs. Evanston needing enhanced tax base partnered with NU to develop tech infrastructure – Park (1985), Incubator (1982).	In 1970s area's leaders advocated the need to diversify its econ base from agriculture to knowledge one. It was envisioned to engage locally trained manpower to develop value added knowledge industry using Virginia Tech resources. As part of the tech infrastructure development VT CRC was established in 1985.	Cognizant of the UW-Madison's R&D strengths, a highly educated workforce coupled with the area's high quality of life, the university officials started deliberating on the development of technology infrastructure during early 1980s. The University Research Park facility was established in 1984.
Technology sectors ● Area's strengths ● Targeted sectors	The region possesses a broad knowledge base with strong info system and manufacturing industry, and developed	The area is high income with educated population and has a strong service sector that replaced smokestack industries	Given Virginia Tech's strengths in electronics, computer & information technologies, and biotech, the	Area's main source of technology, UW-Madison has strength in several technology areas including biotech and info

	infrastructure. Therefore, it emphasizes diversity of technology sectors. However, software, biotech are dominating.	University technology transfer and commercialization in software, biotech and materials is emphasized.	area offered a fairly broad spectrum of technologies. Therefore, there is no special emphasis on a single or a set of sectors other than being high tech.	systems. However, Biotech/ medical and services make the bulk of technologies represented in the area's park.
Existing infrastructure ● Higher education ● Major R&D labs ● Significant firms ● Parks & incubators ● Partnerships ● Risk capital ● Other	The region continues to offer rich knowledge base and highly educated manpower. This comprehensive technology development infrastructure with active private sector, and supportive govt. policies offers an ideal innovation milieu. Seed capital and VC funding is fairly developed. Significant number of univ-related entrepreneurs.	The area offers an extensive R&D-intensive knowledge base, highly trained & educated workforce with entrepreneurial climate. The highly networked technology infrastructure and market driven mixed use Evanston Park and other area incubator facilities offer a sustainable innovation milieu. The venture and seed capital facility is fairly advanced.	This rapidly developing modern innovation pole offers multitude of highly developed infrastructure – parks and newly established incubators. It offers trained manpower and university R&D results. Not only are new firms based on university technology being nurtured, but outside firms are also being solicited through generous relocation packages.	The region is a highly attractive destination for high tech business relocation and new start-ups. With the help of area's utility MG&E a full-blown incubator is also functional. There is already a well developed technology and business development support infrastructure in place. Seed and VC help is being made available.
Overall assessment ● Performance ● Challenges	Strong academic involvement in the region has effectively leveraged its traditionally rich technology base. Specifically, several of RPI's award-winning technology programs provide an innovation milieu that has become the envy of the world. Area's harsh winter climate often poses competitive challenges in capturing and retaining talent.	Area's successful incubation activity has generated a high value added diversified economic base with several newly incubated tech-based firms which provides a potential for future growth in this knowledge-laden environment. The evolving nature of academic–city government partnership especially in Evanston, continues to pose challenges.	Mobilizing its university and regional resources this area has developed a state of the art technology and business development infrastructure that attracts both fledgling tech based firms and those national and foreign firms planning to relocate to the region. Area's seed and risk capital industry is underdeveloped and efforts are underway to improve this.	With the help of university leadership the area has successfully developed a modern and growing technology and business development infrastructure that has earned recognition. The future expansion of the facility and continued reliance of active university involvement has its limitations.

Table 5.3 Comparative overview of the development path of selected Canadian knowledge regions

Development characteristics	The Ottawa Metropolitan area (1.1 million inhabitants)	Quebec City Metropolitan Area (700 000 inhabitants)	Saskatoon Metropolitan Area (240 000 inhabitants)	Calgary Metropolitan Area (1 million inhabitants)
Pre-conditions ● Area's importance ● Skilled manpower ● Higher education ● Private industry ● Other	Seat of Canada's Federal Govt. R&D-intensive region (large govt labs (NRC, CRC), large priv. research lab (BNR, now Nortel). Very small industrial base. Very well educated wealthy labor force (highest % university graduates in country). Two average universities, good engineering schools, one med. school). High quality of life.	Seat of Quebec Provincial gvt. Limited R&D base except for one good research university, a good research hospital and small govt research labs. Moderately well educated population. Mostly government, national defense and services jobs. High quality of life in a French speaking environment. Active tourism industry.	Small city isolated in the Canadian Prairies. Mainly agriculture-based activities. One federal gvt lab (Plant Biotech Institute, ~1955) and a small research university with good agricultural research (plant breeding) and engineering faculty (space science, mining). Moderately well educated population. Low cost of living.	Canada's oil capital, second highest concentration of large firms head offices in Canada (oil firms); moderate R&D base to support oil industry (geomatics, oil research). One good research university. Well educated wealthy population. Good quality of life next to Rocky Mountains (tourism).
Take off ● Restructuring need ● Infrastructure dev. ● Lead actors ● Other milestones	BNR, very large telecom priv. research lab established 1961 to be close to govt labs. Early 1970s, first independent start-ups. Mitel very successful; sold 1973; role model for many other start-ups, some very successful, becoming angels, VCs (A 'home grown region'). In 1983 creation OCRI, bus–govt–univ consortium, to facilitate linkages. 1995, arrival foreign firms, VCs.	In 1983, creation of a research park to support regional economic diversification (a joint gvt–business–university project). Creation of a number of research institutes (INO, CRIQ, Forintec) and of a good number of high-tech start-ups (including spin-offs from the university and INO). Slow but steady technology development.	Decision by provincial gvt in 1977 to create a science park, Innovation Place, as economic development tool for a tech-base community in association with the university. Several public labs were built (1980s, biotech, agri-food, environment), creating a research pole with very strong focus on agricultural biotechnology. The park now houses 115 organizations, 2500 employees.	CR&DA, joint venture university, city, and chamber of commerce created to diversify local economy (1981). Science park (1983), tech incubator (1985). In 1982, creation of NovAtel (cell phones) by Nova Corp and Prov. Gvt. In 1989, research laboratory by TRLabs. Early 90s Nortel opened a large wireless research facility, turning point for local tech development.
Technology sectors ● Area's strengths ● Targeted sectors	Telecom and microelectronics have been the dominant sectors from the early 1970s to the	Information and communication technologies now account for about 4% of local employment	Agricultural biotechnology and nutraceuticals (nutrition and pharmaceuticals) account	The main technology sectors in Calgary (1999) include wireless and telecommunications (10 000

	late 1990s, followed by software products and services and photonics. A nascent life science sector is supported by a science park (1992) and two biotech incubators (2002).	(300 firms). Optics, photonics and geomatics are other important domains (tech transfer from the university and INO). Health sciences is also a growing tech sector (also linked to local teaching hospital).	for about 40% of the local activity. Information technologies (25%) and environment (17%) are the two other leading sectors of this world class highly focused small sized knowledge pole.	jobs), information technologies and software (10 000 employees), geomatics (3500 jobs), IT services (12 000), multimedia (2600), electronic manufacturing (7000).
Existing infrastructure ● Higher education ● Major R&D labs ● Significant firms ● Parks & incubators ● Partnerships ● Risk capital ● Other	Still by far highest R&D intensity in the country with large gvt. labs (CRC, NRC) and private R&D (Nortel, JDS, Alcatel, Mitel). Several large telecom firms, total of 1600 high-tech firms. Two good research universities and excellent bus–govt–univ collaboration. Tech incubation in 2 gvt labs and 2 biotech incubators. Attracts about 30% of all VC investments in Canada.	Good research university and growing – but still limited – focused research base (INO, CRIQ, defense research, Forintec); research park (102 firms, 2500 employees); multimedia (recent) and other business incubators. Limited Venture capital. Low cost region with high quality of life and favourable R&D tax treatment. Affected by provincial politics and language issues.	Innovation Place (science park with 115 organizations, 2500 employees) and the University of Saskatchewan are the two main drivers of local development. Ag-West Biotech (1989), provincial org. dedicated to development of agbiotech industry. Colocation of leading ag-biotech labs (federal, provincial), and TR Labs (1992, information technologies).	Calgary has a superior technical infrastructure, good research base (University of Calgary, Alberta Research Council, Nortel), active research park (72 firms, 1000 jobs), 2 incubators (Technology Entreprise Centre and InnoCentre (recent)) and a number of large telecom/ wireless firms (Telus, Nortel, IBM).
Overall assessment ● Performance ● Challenges	Very successful high-tech development. Now 70 000 high-tech jobs, 1600 firms very high rate of tech start-ups; moved successfully from govt culture to vibrant entrepreneurial culture. Telecom meltdown is main challenge, as is growing tech SMEs to world class status and retaining successful firms in the region.	Successful niche technology business development linked to local university research (optics, health sciences, ITC), but low rate of growth. The region's limited research orientation, small local market, geographic isolation, limited risk capital, transportation issues, and political issues (separatism, language policies) reduce its appeal.	Proactive focused research strategy supported by fed, prov, and local gvts led to the development of a world class research pole with a good number of successful technology start-ups. Limited access to Venture Capital and relative isolation are two of the main challenges for the region.	High growth (population and tech activities), very dynamic, attractive to investors (lower taxes than in other parts of Canada, reasonable costs, high quality of life, entrepreneurial spirit), well educated labor force. Access to venture capital is seen as a problem which could limit the region's growth.

Table 5.4 Comparative overview of the development path of selected Mexican knowledge regions

Development characteristics	Querétaro–Bajío, QB	Medium cities: Cuernavaca Ensenada	Large cities: Monterrey Guadalajara	Mexico City
Pre-conditions ● Area's importance ● Skilled manpower ● Higher education ● Private industry ● Other	Region defined over a continuous area of two industrial corridors. Querétaro located in the crossroads from Mexico City to the North, has changed the industry, from a traditional profile of food and textile (1940–60) to big and modern steel and automotive (1970s). Bajío traditionally based on agriculture also became industrial after the Salamanca oil refinery was established in 1970.	Cuernavaca building up an industrial park (CIVAC since 1960s). After research centers came in, supported the Mexico City area. Research centers with a tradition in fishing and canned fish were created in Ensenada. The proximity to California, US Research institutes (as the Oceanography Scripps Inst. La Jolla, CA) encourages the development of high-tech entrepreneurs.	There is an industrial corridor between Monterrey up to Saltillo. The creation of the ITESM (1943) sponsored by Monterrey's private industry, aid to the creation of the local entrepreneurial culture. In Guadalajara the computing industry was triggered with the location of IBM in 1975. In the 1970s an Industrial Park has been transferred to the Guadalajara University.	Mexico City which accounts for the DF and part of the State of Mexico is the center of the country and it is very industrialized. It has 18% of the Mexican Higher Education (UNAM, IPN, UAM), and 25% of the national university graduates.
Take off ● Restructuring need ● Infrastructure dev. ● Lead actors ● Other milestones	Good education and research infrastructure. UAQ, and other private institutions. UGto has some research faculties as FIMEE in Salamanca applying mechanics, electric and electronics knowledge to agriculture innovations.	In 1975 the creation of the National Lab in electricity (IIE) in Cuernavaca and in 1973 the CICESE oriented to telecoms, physics and oceanic research in Ensenada triggered scientific and tech research activities. Both RCs encourage the creation of spin-offs.	The ITESM has spread nationally with 33 campuses. It began an entrepreneurial program which is now a good source of qualified human resources and of new ideas for the spin-offs housed in ITESM's network of incubators created in 2002.	Big universities developed research groups and centers (1960–80). Five incubators related to universities foster the creation of spin-offs during the 1990s.

Technology sectors ● Area's strengths ● Targeted sectors	The main sectors in the region are Machinery and Agriculture, Transportation Equipment and Chemistry. R&D activities are oriented to metal–mechanical, Chemistry, Transport Roads, Metrology, Optics, Software, and Biotechnology.	The Ensenada high-tech industries are mainly in instruments and electronics, with good research in telecommunications. Cuernavaca industries are software and energy and its research specialty is in biotechnology.	Monterrey research is oriented to Energy, Chemistry, and Materials. Guadalajara has good research on information technologies, Electronics, biotechnology, Machinery and Equipment, and Agriculture.	Mexico City concentrates 50% of Mexico research infrastructure which is quite diversified. Some of the main areas are pharmaceutics, chemistry, instruments and energy.
Existing infrastructure ● Higher education ● Major R&D labs ● Significant firms ● Parks & incubators ● Partnerships ● Risk capital ● Other	The region has a strong technology research base with three national labs, three private research and six public research centers. Two state universities (UGto, UAQ). The incubators and the San Fandila founded in 1994, which went bankrupt. Bajío Incubator (CENIT) has permanently been just a project, partially working.	Cuernavaca has some National Labs: electricity and water resources and other RCs. Ensenada has the CICESE, which is complemented by IPN's CITEDI and UNAM's centers. Cuernavaca has two incubation experiences: Cemit and IETEC. Ensenada has the IEBT incubator.	Both are university regions. Monterrey has RC at the ITESM and at Nuevo León University; in Saltillo is CIQA (chemistry). The ITESM based on its entrepreneurial program is developing a system of incubation. Guadalajara has a successful technology-based incubator and the only one operating science park in Mexico (the 'Belenes Technology Park').	Five incubators related to universities fostered the creation of spin-offs during the 1990s. But only two are in operation in 2002 (IPN's and a UAEM's incubator general business incubation). The main reason for failure was the lack of risk capital.
Overall assessment ● Performance ● Challenges	Strong tech R&D. Good support infrastructure, mainly in industrial parks. Challenges: 1) Learning from incubation and science park failures and 2) To create partnership for VC funds.	Both regions played a pioneering role in creating TB incubator. To develop other specializations based on the outside leverage of Mexico City for Cuernavaca and California US for Ensenada.	Both innovation poles have a good entrepreneurial environment and need to develop Venture Capital funds.	The region's high research density can be leveraged to grow technology research zones and also to complement regional research capacities.

Table 5.5 Key characteristics of US knowledge regions (TIP) with their relative values

		USA			
TIP characteristics	Timing	NY Capital Region	Chicago Metro	New River Valley, VA	Madison, Wisconsin
Key regional actors					
Number of TBF in the region	current	3	3	2	2
Governments and their incentives	past&present	3	2	2	2
Universities and research centers	current	3	3	2	2
The regional context					
Entrepreneurial culture	current	2	2	1.5	2
Qualified manpower	current	3	3	2	3
Quality of life	current	2.5	3	2	3
Lower cost of doing business	current	1	1	2	2
Traditional industrial base	initial	3	3	1	2
Regional infrastructure (telecom, airport, etc)	current	3	3	2	2.5
The innovation process enablers					
Formal incubators	current	3	3	1	2
Science parks	current	3	2	3	3
Financial resources (VC, govt grants, etc)	current	2	2	1	2
Champion entity	past&present	3	3	3	3
Anchor organization	past&present	3	3	3	3

Tech transfer programs	past&present	3	3	3	3
Support services (business, engineering, legal, etc)	past&present	3	3	2	2.5
Networking opportunities	past&present	3	3	2	2
Outcomes					
TBF clusters and innovations	current	3	3	1	2
Total		49.5	48	35.5	43

Notes:
Scales: Qualitative estimate by the authors of the quality and effectiveness of that function within each country: 3 = equal to best in country, 2 = among moderate in country, 1 = among weak in country, 0 = non-existent.

Table 5.6 Key characteristics of Canadian knowledge regions (TIP) with their relative values

		Canada			
TIP characteristics	Timing	Ottawa CMA	Quebec City CMA	Saskatoon CMA	Calgary CMA
Key regional actors					
Number of TBF in the region	current	2	0.5	1	2
Governments and their incentives	past&present	2	2	2	2
Universities and research centers	current	3	2	2	2
The regional context					
Entrepreneurial culture	current	2	0.5	1	3
Qualified manpower	current	3	2	2	3
Quality of life	current	3	3	2	3
Lower cost of doing business	current	2	3	3	2
Traditional industrial base	initial	1	1	1	3
Regional infrastructure (telecom, airport, etc)	current	2	1	1	2
The innovation process enablers					
Formal incubators	current	2	1	0	2
Science parks	current	1	2	3	2
Financial resources (VC, govt grants, etc)	current	2	0	0.5	1
Champion entity	past&present	3	1	2	2
Anchor organization	past&present	3	1	2	3

Tech transfer programs	past&present	2	2	2	2
Support services (business, engineering, legal, etc)	past&present	2	1	2	2
Networking opportunities	past&present	3	1	2	3
Outcomes					
TBF clusters and innovations	current	2	1	2	2
Total		40	25	30.5	41

Notes:
Scales: Qualitative estimate by the authors of the quality and effectiveness of that function within each country: 3 = equal to best in country, 2 = among moderate in country, 1 = among weak in country, 0 = non-existent.

Table 5.7 Key characteristics of Mexican knowledge regions (TIP) with their relative values

		Mexico					
			Middle Cities		Big Cities		
TIP characteristics	Timing	Querétaro–Bajío	Cuernavaca	Ensenada	Monterery	Guadalajara	Mexico City
Key regional actors							
Number of TBF in the region	current	3	2	1	1	2	3
Governments and their incentives	past&present	2	1	1	1	1	1
Universities and research centers	current	2	2	1	2	2	3
The regional context							
Entrepreneurial culture	current	1	0.5	0	3	1	1
Qualified manpower	current	1	1	1	3	2	3
Quality of life	current	2	2	2	1	2	0
Lower cost of doing business	current	2	1	2	2	1	0
Traditional industrial base	initial	1	1	0	3	2	3
Regional infrastructure (telecom, airport, etc)	current	2	1	2	2	2	2
The innovation process enablers							
Formal incubators	current	1	0	0	3	2	2
Science parks	current	0	0	0	0	3	1

Financial resources (VC, govt grants, etc)	current	0	0	0	0	0	0
Champion entity	past&present	1	1	2	3	2	1
Anchor organization	past&present	1	1	1	1	1	0
Tech transfer programs	past&present	0.5	0.5	0.5	0.5	0.5	0.5
Support services (business, engineering, legal, etc)	past&present	1	2	2	3	3	3
Networking opportunities	past&present	1	1	2	3	3	1
Outcomes							
TBF clusters and innovations	current	3	1	2	1	1	3
Total		24.5	18	19.5	32.5	30.5	27.5

Notes:
Scales: Qualitative estimate by the authors of the quality and effectiveness of that function within each country: 3 = equal to best in country, 2 = among moderate in country, 1 = among weak in country, 0 = non-existent.

information can nevertheless be derived from an examination of the characteristics associated with the most successful and less successful regions in each country. The analysis follows the analytical framework of chapter 1 focusing on the key infrastructural elements of a functioning innovation pole laid out in the form of regional characteristics; on organizations, institutions, programs that facilitated the transformation of each region; and on the current status of the region as a TIP (Figures 1.1 and 1.2). Those characteristics and facilitators have been grouped under 4 headings: key regional actors, regional context, innovation process enablers, and outcome (Section 1.5). In some cases, the grouping may seem arbitrary to the extent that some characteristics and organizations have multiple roles in regional development. This is to be expected in dynamic regions with organic milieus and a high level of interaction between organizations, institutions and firms.

5.2.1 Key Regional Actors

Technology-based firms

Some regions are very focused with only one or two clusters of TBFs; other regions have several clusters in complementary or different industries. If the size of a TIP is measured in terms of its number of TBFs, what is the minimum and critical mass for a TIP to be sustainable? What is the optimal size of a TIP? Should a TIP have more than one industry cluster, and, if yes, how many? There is no single answer to these questions, as regional characteristics (geography, infrastructure), cultural norms, sectoral and national differences have an impact on what is acceptable or even feasible. A small number of TBFs may not be sufficient to attract and support the specialized services needed by high-tech firms, be it for legal and intellectual property issues, business and financing issues, or in terms of engineering and technical services, precision machining and molding, specialized testing, prototyping. A small number of TBFs may have too few employees to support the types of schools, retailing and cultural life their employees would expect. On the other hand, too many TBFs in a region may lead to the types of diseconomies experienced in Silicon Valley in the late 1990s due to overcrowding and shortages (of trained manpower, space or energy). The largest and most successful US TIPs (Silicon Valley, Boston, Washington DC) have from 20 000 to 30 000 TBFs; in Canada, those numbers are in the thousands, and in Mexico they are in the hundreds.[4] Clearly, the local infrastructure and local innovation policies and incentives must be adapted to the number of firms present in a region, and updated as numbers change.

Governments and their incentives for research and technology transfer

A number of programs in Canada support technology partnerships, encourage university–industry collaboration, and facilitate technology transfers to and innovation in small and medium enterprises (SMEs). As outlined in section 3.1.2, the Industrial Research Assistance program (IRAP) and its network of industry technology advisors provide technical assistance and R&D funding to SMEs (up to 500 employees) all over the country, the Canadian Technology Network (CTN) helps SMEs to access the services of over 1000 innovation-related service providers, and the Pre-Commercialization Assistance Program offers repayable loans to support commercialization. Similar programs exist at the provincial level. Each province also has an R&D tax credit system (or equivalent), and Quebec's and Ontario's are among the most generous in the country in this respect. Corporate tax rates, including lower rates for SMEs and in particular for technological SMEs, also vary from province to province, providing other regional incentives. In three of the four Canadian TIPS analyzed (Quebec City, Saskatoon and Calgary), the high-tech development process can be linked to a clear motivation by governments (local, provincial, with federal support) working in partnership with local universities to diversify the local economy (this process built directly on the areas of expertise of local universities in Saskatoon and Quebec). In the fourth region (Ottawa), local entrepreneurs took advantage of government research and government purchase of goods and services to initiate the high-tech development process in telecommunications and related fields. In this region, the local and provincial governments decided relatively late to encourage the development of a new technology cluster based on local research (life sciences) by creating a 'life science research park' next to the city's major hospitals, university life science complex and medical school, in an effort to attract more research and private firms to the area.

In the US, as outlined in section 2.1.2, several government programs provide necessary teaming and networking platforms and financial support. These programs include the advanced technology programs (ATP), the manufacturing extension partnership, the partnerships for a new generation of medical technologies, and the United States innovation partnership. The Small Business Innovation Research (SBIR Act of 1982) and the Small Business Technology Transfer Program (STTR, introduced ten years later) provided opportunities for small businesses to get involved in meeting federal R&D needs by providing tangible support for the commercialization of technology (Table 2.3). Numerous technology development programs also exist at the state and regional levels. Some states and regions are more active than others, including Pennsylvania, Georgia, Minnesota, Ohio and North Carolina among others. These most recent so-called 'third-wave' state and regional economic development programs[5] emphasize the regional institutional resources and center around technology development research centers, industry and university partnerships for industrial problem-solving, start-up assistance through incubation centers and small business development centers, project financing and the promotion of networking through regional consortia.

In Mexico, Conacyt began to report directly to the Presidency in 2000 instead of the Ministry of Education as previously. Starting in 2001, its programs were reorganized by sector and type of knowledge: basic science receiving funding from the Presidency and from the Ministry of Education (SEP); applied research and technology development being funded by the Presidency, by other ministries depending on the domain, and also by the states (in the form of mixed funds). Generally, federal funds are directed towards the states with the highest budgets for science and technology, supporting those that are most active to the detriment of the others. A fiscal incentive for undertaking R&D programs was introduced in 2000 to support private sector R&D and commercialization, and since 2003, the AVANCE program encourages partnerships with private funds to support R&D by start-ups. Venture capital for high-tech firms is scarce in spite of the existence of general purpose programs which can be used for value added projects in high-tech (such as NAFIN risk capital and credit). Even if the objectives of these programs are similar to those of other NAFTA countries, their effectiveness is quite low in Mexico, because of the low level of the resources available and the lack of institutional support and control.

Universities and regional research base

All the TIPs analyzed include at least one good research university and most have some public and private research laboratories. However, they vary in terms of research intensity. In this respect, Ottawa in Canada, New York Capital Region and Chicago Metropolitan Area in the US, and Mexico City in Mexico, come out ahead because of the number and size of government and private research laboratories. The concentration of research activities in these primarily urban areas acts as a magnet for high-tech development, providing these knowledge poles with a strong competitive advantage. A recent US Department of Commerce study has concluded that the patenting activity is highly concentrated in a small number of metro areas lending credence to the belief that they offer a much superior knowledge base to exploit (Reamer et al. 2004).[6]

5.2.2 The Regional Context

Entrepreneurial culture

Most US regions studied have had a well developed entrepreneurial culture for a long time. In Canada, the entrepreneurial culture has always been high in Calgary and has grown over time in Ottawa, as the metropolitan area's high-tech industry developed. In Mexico, it is high in Monterrey with its traditional enterprise culture. In that region, ITESM has played a pioneering role in developing entrepreneurial programs within its university system.

Qualified manpower

In the late 1990s, most large knowledge poles in the US and Canada were suffering from shortages of qualified manpower, which sometimes led high-tech firms to move to smaller cities. In 2005, with the high-tech slowdown that followed the 'dot.com bubble', this is no longer the case and most of the TIPs surveyed in Canada and the US have a good supply of trained manpower. This also applies to Mexico's large urban areas, Mexico City, Monterrey and Guadalajara, although the situation is different in the other TIPs – Quetérato–Bajío, Cuernavaca and Ensenada – where the supply of well trained employees is limited.

Quality of life

Some regions are more attractive than others in terms of quality of life, an important factor to attract and retain highly skilled professionals and their families. Three US regions (NY Capital Region, Chicago Metropolitan Area, and Wisconsin-Madison) already offer high quality of life while the fourth region (New River Valley of Virginia, with its attractive climate) has built up new infrastructure and is on the verge of offering most of the amenities contributing to quality of life. In Canada, all of the four regions studied have a very good quality of life. Ottawa and Calgary also have good direct airline links with the United States and the rest of the world, a service judged to be very important by most high-tech entrepreneurs. In Mexico, the medium-sized cities, Cuernavaca and Ensenada, and the Querétaro-Bajío region have a good quality of life, while the bigger cities, especially Mexico City, rank lower on this attribute because of big city problems such as traffic jams, pollution and insecurity.

Cost of doing business

The presence of affordable knowledge workers such as student interns, faculty researchers, affordable housing, reasonably priced utilities, availability of low cost industrial land and raw materials, and low transaction costs added to the attractiveness of some regions more than others. Two Canadian regions (Saskatoon, Quebec City) stand out in their relatively lower overall cost of doing business while the big and medium sized cities of Mexico (Guadalajara, Cuernavaca) and US (Chicago Metro Area, New York Capital Region) were considered least attractive to businesses in this respect. The regions with moderate levels of over all cost of doing business include: Madison Wisconsin and New River Valley in the US, Ottawa and Calgary in Canada, and Querétaro-Bajío, Ensenada, and Monterey in Mexico. Mexico City is unique in this respect with varying levels of generally higher costs in different surburban areas.

Traditional industrial base

The technology clusters that have evolved over time are generally related to their regions' traditional industrial base in three ways:

TIPs in which the technology clusters are derived from or supported by the region's industrial infrastructure. In the US cases of New York Capital Region and Chicago Metropolitan Area, the knowledge regions evolved from an existing manufacturing industrial base. In Canada, Ottawa's high-tech sector has grown around a large telecom company; in Calgary it has been built on research expertise directly related to the town's oil industry and Saskatoon's ag-biotech sector is associated with its agricultural orientation. In Mexico, the Querétaro-Bajío region has a large industrial base, and its clusters are related to the automobile industry and agribusiness as well as the research centers in the region. Mexico City and the large cities of Monterrey and Guadalajara have general support for their technology clusters but lack strong ties to the existing traditional industries, while Cuernavaca has an industrial base that does not promote technology industry. Nonetheless, in these Mexican cases, support is available to develop core technologies based on both industrial and high-tech industries and services.

Some technology clusters play an important role in the regions' industrial diversification. Virginia's River Valley Region in the US is a salient example of the emphasis placed on new technologies as a deliberate regional effort to diversify the agricultural base and establish a knowledge economy. Canada's Ottawa telecommunications cluster developed as a result of local R&D activity and it has changed the region's profile from a government-based service economy to a technology-based economy. Calgary has added wireless and telecommunications to its originally more limited oil orientation. Similarly, optics and photonics, based on local research activities, have diversified Quebec City's economic base from government and traditional wood-related industries into new technology areas. In Ensenada (Mexico), the relative strength in telecommunications that resulted in the development of spin-offs, originated mainly because of the region's research capabilities in spite of its weak local industrial base.

Some TIPs are focused on a single high-tech industry. The University of Wisconsin-Madison is strong in several technology areas, but it focuses particularly on biotech/medical and service technologies. Saskatoon (Canada) has been focusing on agriculture and biotechnology, but it has also added telecommunications and IT to its local high-tech base. Cuernavaca in Mexico is especially strong in biotechnology. In these cases, the region increased its competitive advantage by focusing on one high-tech area. However, this is a risky strategy since reliance on a single industry sector makes a region vulnerable to fluctuating industry cycles.

Regional communications and transportation infrastructure

The presence of a well-developed basic infrastructure in a region in the form of modern communication and transportation facilities including telecommunication networks, airports, roads and waterways provide necessary links within and outside the regional communities. Among the 12 cases (14 TIPs) studied, two US regions (New York Capital Region, Chicago Metro Area) have the most advanced infrastructure facilities, while the other two regions (Madison-Wisconsin, New River Valley Area) offer solid infrastructures which are continuously improving. In Canada, Ottawa and Calgary also have solid communications and transportation infrastructures with direct air links to major US cities and TIPs, as is the case with most Mexican TIPs. Quebec City and Saskatoon in Canada and Cuernevaca in Mexico tend to be somewhat isolated with only a limited number of direct air-links to major US destinations.

5.2.3 The Innovation Process Enablers

Science parks and incubators

The relatively new economic development tools, science park and technology incubators, have been pioneered in the US where they are now a well-developed industry. The four US cases included in this study have very well established state-of-the-art science park facilities, and two also host reputable technology incubation centers. Canada also has some very successful world-class science parks and technology incubators,[7] but the performance of the parks and incubators in the four regions analyzed is mixed. Saskatoon, Calgary and Quebec City have well established successful parks, while Ottawa's is only beginning to grow after a long gestation period, and only three regions, Calgary and Quebec (for many years) and Ottawa (recent creations), have incubators with varied levels of success. The low priority given to science parks and incubators in Mexico and/or their poor performance (except for Guadalajara) show a big gap with the US and Canada.

Venture capital

A number of US regions, including the four cases in the study, have relatively well developed venture capital (VC) industries (the rural New Valley Region is however making concerted efforts to improve), although the amounts of funds available in 2004 are back at pre-1998 levels after the highs of 1999–2001. In Canada, there is also a relatively well developed VC industry that grew with the growth of the regions, venture capitalists being attracted by success (best developed in Ottawa, followed by Calgary, Saskatoon and then Quebec City). In the US and Canada, governments provide some financial assistance for SMEs for the development and commercialization of new technology and services, slightly reducing the risk to private investors and making their investments more appealing. In contrast, there is a clear lack of venture capital in all the Mexican regions. It would be advisable for the government of Mexico to follow the US and Canadian examples by developing public policies conducive to the creation and development of private venture capital funds.

Networking, partnerships, and champions

Networking is a very important characteristic of successful TIPs, as it enables knowledge to be created, shared and used. Of particular importance to the innovation process are (1) networking among scientists, engineers and researchers in universities and public and private research laboratories, and (2) networking between researchers and the business sector (industry, business services and venture capitalists).

In each country, the extent of involvement of researchers and scientists, universities, large and small firms differs from case to case. In all of the US cases studied, partnership arrangements involving research universities, governments and the private sector played a prominent role. As described in Chapter 2 there have been numerous committed individuals and other champion entities within each of these three sectors that have spearheaded the establishment of incubation mechanisms. In Canada, Ottawa's and Calgary's success is due in no small part to the level of networking and the quality of the linkages developed between firms, universities and government organizations thanks to the proactive work of local university–industry–government consortia that have been strong champions of local advanced technology cluster development (OCRI in Ottawa, CR&DA (now CTI) in Calgary). Ag-West Biotech (and SOCO) have played a similar role in Saskatoon.

Technological change and innovation come from new market-based alliances between industry and science, especially for the commercialization of research through spin-off companies and the licensing of intellectual property. In all the TIPs surveyed, there are partnerships between local universities and local firms (often SMEs looking for short-term, problem-solving capabilities), either directly or through intermediaries (local consortia, centers of excellence). In the US's NY Capital Region and Chicago Metropolitan Area, in Ottawa and Calgary in Canada and in Guadalajara in Mexico, these alliances also involve multinational enterprises and the local universities.

In Mexico, some universities and research centers have well established relationships with industry either in the form of research linkages or delivery of services, or both.[8] Networking is better developed in the big cities of Monterrey and Guadalajara, each of which has a well established industrial base and proactive research universities – respectively the ITESM and the University of Guadalajara (UdG). ITESM's networking comes from its entrepreneurship programs and also from its advanced research activities in some focused areas (solar energy, manufacturing, technical services and environmental assessment). UdG established the university industrial park 'los Belenes' to promote innovation activities, and built a successful incubator and a science park.

Anchor organizations

One of the other conditions that prompted the number of local TBFs to start growing was the presence of one or more large research/manufacturing firms with a large number of engineers and scientists that could operate as an 'anchor organization'. Such an anchor organization makes the region appealing to other engineers and scientists in similar or complementary fields, is a potential client for SMEs, and can be a source of potential spin-offs. In the New York Capital Region, General Electric contributed in a significant way through its spin-off activity. In Ottawa, Nortel has been that key anchor firm, followed by Mitel and Newbridge. In Calgary, NovAtel, followed by Nortel, have had the same role. IBM in Guadalajara has been the main anchor firm. It is worth noting that the rate of development in Ottawa and Calgary (with their large private research/manufacturing firms as anchor organizations) has been higher than that in Saskatoon and Quebec, where research laboratories have served as anchor organizations, and one or two successful local start-ups had been good role models but were probably too small to serve really as 'anchor firms'. On the other hand, in the Northwestern Evanston park region the BIRL and ARCH research lab complexes, though well intended, have so far yielded limited results in creating NTBFs.

5.2.4 Outcome: Regional Development Path, Current Status and Sustainability

The current status of each TIP analyzed in this book is a result of the evolution, spanning over the past 20 or 30 years, of regions that have been transformed by the development of new technology clusters. Each TIP has followed a unique development path in line with its initial state (pre-conditions), the regional context, local enablers and champions, and a host of national and international factors. This section delves into the factors that made the regions start or accelerate their transformation into modern knowledge regions, the champions that made it happen, the institutions and other factors that seem to be related to their success, and also the elements that lead to medium-term and long-term sustainability.

A general precondition for the development of all the knowledge regions surveyed has been the presence of at least one good research university, source of highly trained manpower and sound research, a necessary but not a sufficient condition for growth (Doutriaux, 2003). In several cases, Calgary and Saskatoon for example, the number of TBFs in a specific domain (external firms moving in or local start-ups) started to grow significantly only after the establishment in the region of several public (federal, state, provincial) or private research laboratories creating a critical mass in research in that domain. And as stated earlier, in most cases, a large research/manufacturing firm with many engineers and scientists served as 'anchor organization', making the region attractive to other scientists and engineers, a source of contracts for products and services for smaller firms, and a breeding ground for technology spin-offs.

One of the main findings of this study is the importance of the level of cooperation between the three key regional actors – higher education and research laboratories, private industry, and governments (federal, provincial/state, local). A high level of cooperation among the three actors is more important than which one operates as the primary mover behind development. The analysis also shows clearly the importance of a proactive individual champion, or better, an institutional champion, development organization or consortium, to promote development, ensure continuity, and relentlessly encourage cooperation between the regional actors. US cases show the academic sector's lead role in the New York Capital Region, New River Valley Region and the Madison Metropolitan areas. In Canada, provincial governments seem to have taken the lead role initially in several of the areas surveyed but success seems to be associated with the creation of dynamic public organizations (as Ag-West Biotech in Saskatoon) or consortia of university, government and industry (CR&DA in Calgary, OCRI in Ottawa).

Universities, governments and firms have played proactive roles in the development of innovative regions in Mexico, most frequently in the form of research centers. In Querétaro–Bajío, Cuernavaca, Ensenada and Mexico City, research centers have been the most ardent promoters of innovation and spin-offs (CIATEQ, II Electricas, CICESE, I-Ingenieria-UNAM, respectively). In addition to the federal government programs mentioned earlier, some state governments are also active in this matter as Querétaro, recently in Nuevo Leon. In some cases, development has also been promoted by municipal governments, as at San Pedro Garza Garcia in Nuevo Leon, but these initiatives did not lead to the development of active TIPs, as the other key actors did not materialize. Most of the emerging Mexican TIPs are therefore stagnant.[9]

Some regions – particularly New York Capital region, Chicago Metro Area, and Monterrey – already had well established traditional industrial bases before their new advanced technology clusters developed and expanded. They also had access to the engineering, technical, business and legal services found in large industrial cities, as well as the financial institutions and resources used by established businesses. In many of the other areas, these services and financial resources, especially risk capital, were either non-existent or available on a very limited scale. In several of these areas (Ottawa, Calgary, Madison-Wisconsin), these important resources started to develop locally once a minimal critical mass had been reached, and gradually developed as the areas continued to grow. In the Ottawa area, for example, it was not until the mid- or late 1980s that precision molding, precision machining and prototyping firms found enough local business to start operating locally. It was not until the mid-1990s that the number of venture capital firms and intellectual property law firms started to grow significantly. One complaint from entrepreneurs interviewed in Quebec City was that they had to go to Montreal, 250 kilometers away, for some of their specialized services, Quebec not having reached critical mass for the particular service needed. Services are generally market-driven, and whether they are technical, financial or legal, they tend to lag behind local needs, putting local entrepreneurs at a disadvantage. It is therefore advantageous for local development organizations and local governments to offer appropriate incentives to make them available earlier than would be justified by pure market forces. This is what is commonly done in brick-and-mortar incubators and science parks.

Medium term and long term sustainability come with critical mass and diversification:

- critical mass is necessary in order to have a vibrant high-technology community that is attractive to highly skilled professionals, a market large enough to support the various technical and business services required by TBFs, and the schools, cultural life and other amenities that contribute to quality of life.
- diversification reduces the area's dependency on technology and sectoral cycles and provides scientists and engineers affected by a downturn in their specific cluster with new local employment opportunities

5.2.5 Concluding Comments on Building Up TIPs

The 12 cases (covering 14 TIPs in three countries) show that knowledge clusters can develop successfully in various types of environments provided there is a solid local research base and a supportive local culture. Not surprisingly, the characteristics of the most successful TIPs are as described in the literature: a solid focused research base; large and small TBFs with at least one large anchor organization; a good supply of skilled manpower and related institutions (universities, post-secondary technical colleges); an entrepreneurial culture that encourages networking, partnerships, cooperation; active cooperation between university and research organizations, governments and industry; supportive government programs; technical, legal and business services including venture capital and other sources of financing; solid regional infrastructure (telecommunications, transportation); and high quality of life.

The development cycle of the TIPs in our sample suggests that it is important to:

- Start out by building a solid research base or reinforcing the existing one, or both, in order to reach the critical mass needed for national and international visibility and to become attractive to highly skilled researchers and scientists. This must include a research university, source of knowledge and skilled manpower, and preferably major public and/or private research laboratories. This solid research base seems to be a prerequisite for the growth of a number of TBFs in the region, whether home grown or migrating from other regions. Such a research base may already be present before the beginning of the development of new advanced technology clusters, as is the case of regions with a well established industrial base (Chicago Metro, New York Capital Region, Monterrey, Guadalajara), or with solid government research (Ottawa), or it may be developed, by attracting or growing world-class public and private research centers and laboratories (New River Valley, Saskatoon, Quebec, Ensenada).
- Develop conditions that will make the region attractive to at least one large private research/manufacturing firm, which will act as an anchor organization. The presence of a major TBF in the region provides international visibility and many high-tech jobs. The anchor firm itself is a potential client for local SMEs and a potential source of spin-offs. The development of those conditions may include working with local institutions of higher education to ensure a steady supply of appropriately trained manpower; government programs designed to facilitate technology transfer; measures conceived to attract the engineering, technical, business and legal services needed by TBFs in the region, including risk capital and other sources of financing, which might not be available immediately if they were left to local market forces. They also require a good physical infrastructure (telecommunications, airport, serviced land).

- Have a local champion, individual or organization (preferably an industry–university–government partnership) dedicated to the development of the area as a knowledge region, motivating and ensuring full cooperation between the key partners (industry, university and governments), and providing continuity in the development process.
- Develop an entrepreneurial culture and promote networking and cooperation within and between research places (universities and research centers), industry, services (including venture capitalists and business services) through appropriate events and activities.

Some regions with a well established industrial infrastructure may already have most of these attributes, including large firms, extensive research activities, active industry and trade associations, extensive business and technical services. Other regions may have to develop these attributes, generally starting by focusing on one type of activity (biotechnology in Madison Wisconsin, ag-biotech in Saskatoon, opto-electronics in Quebec, telecommunications in Ensenada and biotechnology in Cuernavaca). Formal incubation mechanisms (science parks, brick-and-mortar incubators) may also be helpful by providing a focal point for technology development and startup activity and by making the region more attractive to research organizations, entrepreneurs and TBFs.

5.3 INSTITUTIONS AND MECHANISMS FOR NURTURING INNOVATION

This section focuses on two types of mechanisms used to nurture innovation in our 12 regions: technology incubators and science parks. The purpose is to identify best practices that could eventually be considered, with some adjustments depending on local conditions, to develop institutions and policies for innovation.

5.3.1 US Science Parks and Incubators

As stated earlier, the US is a leader among NAFTA countries in terms of science parks and incubators (section 2.1, 2.2). This section will highlight the role these mechanisms have played in the development of the four US knowledge regions included in this study. All four regions have relatively well established science parks and two have highly successful technology incubation centers. All are good examples of successful incubation mechanisms. Table 5.8 provides an overview of the characteristics of the four US cases (for details see Chapter 2).[10]

Table 5.8 Comparative overview of the benchmarking characteristics of selected US science parks and their incubators

Incubation Mechanism Assessment	RPI Technology Park	Northwestern Univ./ Evanston Research Park	Virginia Tech Corporate Research Center	University Research Park, Wisconsin
Facility background ● Type of mechanisms ● Sponsors, year established ● Tenant firms and employees ● Other	A suburban park with a solid incubator facility, estab. in 1982 by RPI. About 80 firms and 2000 employees. Broad range of university and area technology focus.	A suburban mixed-use facility with incubator, estab. in 1985 by NU with city's participation. About 62 firms, 1000 employees. Software, biotech, materials focus.	A rural park with a virtual but developing incubator facility. Estab. in 1985 by VT. About 104 firms and 1800 employees. Broad range of university technology.	A suburban park with maturing incubator. Estab. in 1984 by UW. About 110 firms and 2300 employees. Focus on biotech and other areas of univ. technology.
Performance outcome ● Program sustainability, growth ● Firms survival, growth, innov. ● Contribution to sponsor's mission ● Regional impacts	Sustained, grew over the past 2+ decades. High firm survival rate with substantial innovations and involvement of univ. technology and alumni.	After 15 yrs of sustained growth, park transformed into mixed use. High firm survival rate at incubator with successful firms. Entrepreneurship training for students/ faculty.	Experienced fast growth in facilities and tenants since inception. Salient efforts in university tech tr. and entrepr. training. Steady tenants generating substantial innovations.	Sustained growth in park over the past 2 decades. Significant efforts in univ tech. tr. through WARF. Created successful firms and partnered with private sector to help incubate new firms.
Management policies & their effectiveness ● Goals, structure, governance	Governed by RPI board of trustees. Funded by RPI, state funds. Strict entry and flexible exit criteria. Univ.	Park is for profit-corp. governed jointly by university and city. Incubator not-for-profit	Park is not-for-profit subsidiary of VT foundation. Governed by board with regional reps.	Park is organized as 2 not-for-profit corps. under univ. trustees. Strict entry and flexible exit criteria. Serves

● Financing and capitalization ● Operational policies ● Target markets	training & tech tr., reg'l economic development objectives achieved.	with strict entry and flexible exit criteria. Tech trans, training, area economic development objectives achieved.	Strict entry and flexible exit criteria. Serves as a focal point in economic development of the region.	as a focal point in mobilizing univ. developed technology for regional development.
Services and their value added ● Shared services ● Networking support ● University-related services ● Tenant perceptions ● Other	All shared services, business and tech development services, seed cap. and VC access, info databases, tech tr., access to univ. resources. Tenant satisfied with support & avail. of RPI resources.	Most shared services, business and tech development services, easy seed cap. and VC access, info databases, tech tr., access to univ. resources. Tenant generally satisfied with support & avail. of NU resources.	Limited shared park services including tech development services, efforts to provide seed cap. and VC access, info databases, tech tr., access to univ. resources. Incubator service facilities being developed.	Increasingly developed park and incubator shared services, business and tech development services, easy seed cap. and VC access, info databases, tech tr., access to univ. resources. Tenant satisfied with support & avail. of resources.
Overall assessment ● A process view and performance challenges	Award-winning successful facility serving as a model. Continued efforts to involve stakeholders, and attract talent & technology.	The facility has an award-winning successful incubator. With the mixed use approach, efforts are underway to maintain univ. tech tr. focus.	A modern and growing park facility with virtual incubation component. Sustained efforts to capitalize on university tech and developing area innovation infrastructure.	A state of the art park with a unique incubation operation. Developing a model tech entrepreneurship infrastructure based on university's R&D strengths.

Situated in the New York Capital Region, Rensselaer Technology Park (RTP) is a nationally recognized TIP. RTP is a suburban science park established by the RPI University in 1982 and equipped with an incubation center. With an all-technology focus, the park boasts of housing over 80 firms and 2000 employees. With the involvement of university-related entrepreneurs and technology, the facility has experienced steady growth in the past two decades, yielding substantial innovations and a high firm survival rate. A committee of the RPI University's board of trustees governs the park. Over the years the RPI has funded RTP with the help of occasional state funds. The park follows a strict entry policy based on its objectives of supporting university technology commercialization and the area's economic development by nurturing technology-oriented firms, while the exit policy is kept flexible. The park provides shared physical space and office services, business and R&D support, access to seed and venture capital, and makes university resources available to its tenant firms – most of the tenant respondents were generally satisfied with the provision of these services. RTP is an award-winning facility that is considered as a successful science park model. In its strategic plan the park anticipates sustainability and continues to encourage better stakeholder involvement to attract entrepreneurial talent and new technologies.

The Northwestern University/Evanston Research Park is another reputable innovation pole situated in Evanston, a northwestern suburb of the Chicago Metropolitan Area. The park was established in 1985 by Northwestern University in partnership with the city of Evanston. It houses over 60 tenant firms and 1000 employees. The focus is on software, biotechnology and material sciences in which the university has a considerable strength. After 15 years of sustained growth, the park has recently transformed into a mixed-use business and technology park facility. There is a highly successful award-winning incubation program with a high tenant firm survival rate. The park provides an entrepreneurial training ground for the university's students, alumni and faculty. It is a for-profit corporation governed jointly by the university and the city of Evanston. The incubator part is not-for-profit and has strict entry and flexible exit policies that aim to support technology transfer and the area's economic development. The park provides shared space and most shared office services, business services and R&D support, developed seed and VC provision as well as referral services, and access to university resources; most tenants surveyed reported that they were satisfied with the level of park services. With its new mixed use emphasis, the park is trying to maintain its technology emphasis, and efforts are underway to transform the region around the park into a fully networked modern technopole.

Virginia Tech Corporate Research Center is a university related science park situated in Virginia's New River Valley Region. A rural park facility with a virtual but developing incubator operation, it was established in 1985 by

Virginia Tech. There are more than 120 tenant organizations and 1800 employees, and a broad range of university technologies are represented in the park. The park has experienced steady growth in physical facilities and tenants since its inception and it offers state-of-the-art infrastructure in technology transfer and entrepreneurial training that generates substantial innovations. The park is a not-for-profit subsidiary of the Virginia Tech Foundation, which is governed by a board with broad regional representation and serves as a focal point in economic development by incubating and capturing technology-oriented firms in this rural region of the state of Virginia. Due to its technology emphasis, entry policy is restricted toward those firms that can benefit from regional technology resources. The park's exit policy is kept flexible due to the absence of a shared physical incubator space. With its expanding park, and incubation advisory and shared services, the facility is making sustained efforts to tap university technology by modernizing the area's technology infrastructure through broadband connectivity and by leveraging university resources. The VC and seed capital availability is limited and there are conscious efforts to improve this by colocating financial institutions on the park facility.

University Research Park in Wisconsin-Madison is a suburban park with a maturing incubator facility. Established in 1984 by the University of Wisconsin at Madison, the park has over 110 tenant organizations and 2300 employees. The focus is on university-developed technologies managed by the Wisconsin Alumni Research Foundation (WARF), specifically in the biotechnology area. The park has been successful at incubating several new firms in partnership with the local private sector, which is also funding the incubator operation. The park is organized as two not-for-profit corporations, one for real estate operation and the other to run its Science Center. A committee of university trustees with local representation governs it. The park provides shared space and office services to its client firms, as well as most business development services, including VC and financial referrals and technology support services from university resources. A survey of the clients showed that they were generally satisfied with the provision of these services. URP is a state-of-the-art award-winning facility with a unique incubation experience in longer-term public-private partnership. Over the years the park has been instrumental in developing a university-led model of entrepreneurial infrastructure to support technology oriented innovative firms in the region.

In summary, these US cases show that science park and incubator mechanisms have been instrumental in developing successful technology innovation poles in diverse regional settings. These case studies do provide some insight for those interested in establishing this type of technology development mechanisms in their own regions. Several lessons can be learned from these US cases in terms of leveraging regional resources to successfully nurture TBFs through the establishment of science parks and incubators:

a. The presence of a knowledge base in the form of research universities and/or major R&D institutions serves to generate new knowledge and trained manpower. In all four cases the respective universities and their R&D centers played this crucial role.
b. Successful partnership of the university and/or R&D institution with public and private sector institutions enables the necessary mobilization of resources for the pursuit of entrepreneurial ventures such as access to information, shared governance, tapping university resources, availability of private venture capital. The above four cases offer salient examples of university–industry–government partnerships, in most cases allowing ample sharing of resources.
c. The presence of champions at various levels of the regional system provides the necessary motivation to initiate and persistently pursue the development of projects such as the establishments of science parks and incubators. The existence of champions at various levels is evidenced in all the four cases (see Chapter 2 for details).
d. Once established, parks and incubators must secure the funding and professional management necessary for their operations. They must also devise an entry policy that targets certain technologies and allows a flexible exit in line with the facility objectives. They must also provide shared space, office services and business and technology support that contribute to tenant firms' survival and growth. In the US cases, all four facilities have endeavored to secure necessary funding and professional management to run their facilities, and three of them provide all of the services. Only the Virginia Tech-CRC does not have a formal incubator, though there are plans to establish one by 2005.
e. The performance of science parks and incubators as innovation mechanisms needs to be assessed through their sustainability and growth, the success of their tenant firms, the overall contribution to the sponsor's mission, and their regional impacts (for details see Mian, 1997). In all of the four cases, the science parks and incubators reported carrying out performance assessment activities through benchmarking with their peers.

5.3.2 Canadian Science Parks and Incubators

Each of the four Canadian regions analyzed in this project has a science park, as only regions with this feature were selected for study. All except Saskatoon also had one or more incubators in the 1980s or 1990s. Since 2000, some regions have developed additional incubators (Chapter 3). The evolution of these incubating facilities has been described in detail in sections 3.2.1 to 3.2.4 and will only be summarized here.

Science parks

The primary objective of the Quebec, Saskatoon and Calgary science parks, developed in the 1980s, was technologically-based local economic development. No specific sectoral focus was indicated, the goal being to leverage local research activities (university, public laboratories, local industry) and facilitate their commercialization (Table 5.9). The result is that these parks pursue activities generally in tune with each region's research base: optics and IT in Quebec, agricultural biotechnology and IT in Saskatoon; telecommunications, wireless, geomatics and oil research in Calgary. These three science parks are quite successful, as they were able to attract many public and private research laboratories in their early years, followed by high-tech firms, high-tech start-ups, and services organizations, to reach reasonable levels of development within a decade of their creation. Ottawa's situation is different, as its research park barely grew during its first ten years. It was created in 1992 to support the development of a nascent life science sector, a new sector in an R&D-intensive region in telecommunications, photonics, microelectronic and software which had developed since the mid-1970s without the help of any formal technology incubator or research park. Even though the local universities (all with research focus) were involved in the creation of these science parks and even though they have representatives on their board of directors, they are only indirectly linked to park operations.

Innovation Place, an internationally recognized award-winning science park in Saskatoon, has been a key factor in the region's ability to attract, first, specialized federal and provincial research laboratories, and soon thereafter, private technology-intensive firms and technology and business service firms. In doing so, it has helped the region to become a world-class knowledge pole in ag-biotech. Its success comes from the excellent collaboration between the federal and provincial governments and their decision to concentrate federal and provincial ag-biotech research activities in the park to create a highly focused knowledge pole with the critical mass needed to become attractive to external organizations. This is an excellent example in which a science park has generated regional development in an isolated medium-sized community with the help of an excellent research university. With its mix of research institutes, private high-tech firms, and private technology and business service firms, the park has become a dynamic organic environment, supportive of research and innovation, of public–private–university partnerships and networking, the center of such activities in town, the place to be for science-based activities.

Table 5.9 Science parks in the four Canadian regions under study

Science Park	Ottawa Life Science Technology Park	Quebec Metro High-Tech Park	Innovation Place	University of Calgary Research Park
Objective, date created, primary movers	Urban park (9ha) created in 1992 next to local hospitals and School of Medicine to support the development of a health sciences cluster in the region. Moved initially by city, province and university; now managed by consortium university, research centers, local firms, government.	Suburban park (135 ha) created in 1983 for regional economic diversification (research and TBFs) by local (25% op. costs), provincial, federal governments (30% each until 2002). Large board of directors, university well represented.	Urban park (48ha) created by Province and University in 1977 on university land to support the development of a technologically based community in association with the university. University well represented on board of directors. Managed by provincial entity.	Urban park (50ha) created in 1983 by city, Chamber of Commerce and university on government land next to university to foster pure and applied research and help develop tech business. Managed by non-profit entity created by three founders; large board of directors with university and research representation (30%), and large industry representation.
Sectoral orientation	Strictly focused on health sciences and biotech.	No strong sectoral focus, research based (optoelectronics, IT, telecom, new materials, nordicity), and traditional (forestry).	Main focus on ag-biotech, but also IT, electronics and related support services.	No official sectoral orientation; mainly IT, telecom, software.
Number of firms and buildings	One multi-tenant building (8 tenants) built 1992, and one biotech incubator built 2002.	100 organizations (14 buildings) with over 2500 employees in 2001.	115 organizations, 2200 persons in 23 buildings.	Over 3000 persons in 7 buildings (total over 1.2 million sq feet).
Services available in the park	Usual services offered in a multi-tenant building, including reception and conference space, meeting rooms, etc. Networking	Business services to SMEs provided by park administration: technology assessment, business planning, export promotion, marketing,	Most business services available through park tenants, business, accounting, legal, IP issues, etc. Networking activities and other tenant-	Some business services available in Discovery Place, a large multi-tenant and incubation building in the park. Access also to university

	opportunities, workshops, conference series.	etc.; networking activities and workshops, seminars.	related events, workshops, seminars; access to fully equipped laboratories, etc.	facilities.
Techology incubator?	One biotech incubator built 2002; until then mostly multi-tenant building.	CREDEQ until the late 1980s. No incubator currently.	No.	Technology Enterprise Centre at Discovery Place.
Comparative advantages (according to some tenants) and regional impact	Prime urban location for health research activities, next to hospitals and medical school; focal point for nascent health science cluster.	High concentration of research; regional pole in optics; excellent location; prestigious address.	High concentration of research; availability of all business services (by park tenants); a very good address; high-tech nucleus of region.	High concentration of research, in domains representative of local high-tech base; closeness to university, good support services; a very good address.
Overall assessment and future challenges	First ten years very slow; with the opening of the Ottawa Biotechology Incubation Centre (OBIC) in 2002, the park finally seemed to start growing; it has been the catalyst for the growing biotech cluster in Ottawa.	Started well thanks to federal and provincial funding; went through difficult time in mid–late 1990s with the phasing-out of public funding; has been an important focal point for local high-tech development; now in danger of losing its science orientation and becoming only a nice industrial real-estate development.	Very successful; has been named one of the 20 best 'University-Related Parks' in the world in 1990; key factor in the development of the region as a world class ag-biotech center; current challenge is to attract venture capital to sustain the growth of its small firms; may suffer from current questioning on genetically modified plants.	Quite successful as an up-scale high-tech real-estate development, with well located land for R&D activities; has definitely contributed to the high-tech development of the region; unclear if it has contributed to the creation of many high-tech spin-offs (except possibly at Discovery Place) because of an apparent lack of synergy between tenants, a lack of interactions between the major tenants (who have separate buildings).

Real estate has probably been one of the biggest advantages of the University of Calgary Research Park, which has provided lease-able well-serviced land in a nice location next to the university and relatively close to the center of town. As such, the park has been a very useful addition to the local technology infrastructure. It houses Discovery Place, a large multi-tenant/incubator building, as well as research centers belonging to several large corporations and does not have a specific sectoral focus. It does not seem to have been able to develop the organic culture that facilitates and encourages informal networking within and between its tenant research organizations, possibly because of its urban location. In spite of its proximity to the University of Calgary, one of its founders, few university spin-offs are located in the park. The University of Calgary Research Park is certainly a nice place to be for research-based organizations, but it is not the only place in town for this type of activity.

The Quebec Metro High-Tech Park started well, under the leadership of GATIQ, the 'Groupe d'action pour l'avancement technologique et industriel de la région de Québec',[11] a university–industry–government consortium created to develop a solid high-tech base in the region. The park had no official sectoral focus, attracting quickly several public research laboratories that were followed by high-tech firms, spin-offs from university and research labs, and service organizations. Those early years were facilitated by generous funding from the federal and provincial governments covering the park's operating costs (a significant subsidy to tenants) and made successful by solid cooperation between the various levels of government, the business community and the university. In the mid-1990s, however, park development stalled as a result of the phasing out of subsidies, lack of leadership and reduced cooperation between local stakeholders. The park has now been privatized and its operations are financed by the sale of land to technology-based firms and research organizations. The Quebec Metro High Tech Park is considered to be a prestigious address in the Quebec City region. It spite of its name and the benefits of colocation with other technology-based organizations, the availability of low cost serviced land in other areas reduces the appeal of locating there.

In 2003, after an early change of governance and around ten years of slow operations (with only one multi-tenant building), the Ottawa Life Science Technology Park is starting to develop rapidly with a new biotech incubator and other buildings under construction.[12] After a long gestation period, the life science-biotech sector in the region is starting to grow, under the leadership of the Ottawa Life Science Council (OLSC), a public–private partnership representing local life-science interests including government research laboratories, local hospitals, the University of Ottawa and private firms. OLSC is a key driver in the development of a life-science sector in the region, and the park, located next to the area hospitals and the University of Ottawa's Faculties of Medicine and Health Sciences, has been a focal point in technology development.

Generalizations are difficult to make based on the observation of only four science parks with different objectives located in different environments. Nevertheless, it may be useful to note that our analysis of those four Canadian cases shows that successful parks seem to:

- require the solid cooperation of local and other levels of government, local industry, and universities;
- need a solid research base (several public and private research laboratories) to create a critical mass attractive to other research organizations and high-tech firms and services;
- benefit from the leadership of a champion, generally a local consortium of public and private organizations, firms, universities (OLSC in Ottawa, CR&DA (now CTI) in Calgary);
- provide value based on the intensity of knowledge activities, networking opportunities, proximity to research laboratories or a university, availability of services, rather than monetary incentives (low rent or low fees).

Technology incubators

As noted previously, Canada does have some award-winning technology incubators,[13] but, until recently, the number of formal technology incubators in the country was quite limited. It is only in the past four or five years that a few universities have opened technology incubators to assist their researchers in the spin-off process, and the number of incubators in research parks remains very small (Doutriaux, 1998).

In our four Canadian cases, no relationship was observed between TIP success and formal technology incubation and, except for the science parks described previously, formal technology incubation is not very well developed:

- none of the very good research universities in our TIPs had a technology incubator at the time of the study;
- two universities were involved in the early years of two of the incubators covered by the study (Quebec and Calgary), but their involvement has decreased over time;
- there is no formal technology incubator in Saskatoon, a successful TIP with an excellent research park;
- there were no technology incubators in the Ottawa area until 1994, when the region was already a well established TIP.

As outlined below, the incubators in our four Canadian regions have very different objectives and profiles.

In Ottawa, the objectives of two federal government research laboratories, the CRC Innovation Centre, created in 1994, and the NRC Industrial Partnership Facility, created in 1998, were to facilitate technology transfer between scientists and industry and to provide direct assistance to the research teams of existing high-tech firms, especially SMEs and start-ups. These two facilities are very efficient and perform successful technology transfer operations, providing their tenants with direct access to labs and technical expertise. Tenants must be sponsored by a government lab researcher (research mentor) and are admitted for up to three years. The incubators' primary objective is technology transfer rather than the nurturing of start-ups. CRC initially offered some of the business assistance services typically found in technology/business incubators, but these services have been discontinued for lack of demand by tenants. Indeed, such services are readily available from many other sources in the region, in particular the Ottawa Centre for Research and Innovation (OCRI), a proactive consortium of high-tech businesses, universities and government entities, and its Entrepreneurship Centre, a virtual incubator offering courses and workshops, counseling, mentorship, access to 'angels' and venture capital, and so on. The main advantages noted by tenants of CRC and NRC include direct access to labs and researchers, secure facilities with common spaces, on-site industrial technology advisors (an NRC SME support program), business counseling and networking advice if needed. Rent is above market price, but this is not considered an issue by the tenants who were interviewed, given their ease of access to CRC and NRC facilities. A third incubator opened its doors in Ottawa in 2002, the Ottawa Biotechnology Incubation Center (OBIC), an addition to the health science and biotechnology infrastructure developed to support the development of a health science cluster in Ottawa.

In Quebec, CREDEQ was created in 1989 to provide assistance to small tech-based firms and start-ups through the provision of subsidized space and business services. This incubator has no specific sectoral focus. It started well, admitting tenants with good business plans for periods of up to three years. In 1999, it had 52 graduates including at least 41 survivors with combined sales of C$52 million and combined employment of 490 (1 to 150 persons per firm). However, it has not met expectations over time and, in the late 1990s, had problems finding good tenants who were admitted for indeterminate durations. Initially located in the Quebec Metro High-Tech Park and in two buildings several kilometers away donated by the municipality, it lost its research/technology orientation when it closed its offices in the park, its main remaining competitive advantage being subsidized rent and services. One local entrepreneur, interviewed in 2000, said that, even if he did consider joining the incubator early in the development of his firm, he decided against it because he felt that it was very important to be in the park, where the 'action' was. In 2000, when its external funding ended, CREDEQ had to start charging real rents and fees for services, and lost most of its tenants. Its lack of a sectoral focus that could create a community of common interest for start-ups and its location away from R&D facilities did not provide it with the competitive advantage that could have justified paying for services readily available from a number of virtual incubators or other types of non-profit or public organizations.

In Calgary, the Technology Enterprise Centre (TEC), located in the University of Calgary Research Park and created in 1985 by the local development organization also responsible for the park, operates as an up-scale multi-tenant building/incubator with leases of variable duration, some up to 10 years. A number of local firms located outside the park are also affiliated with TEC and have access to its services, workshops and networking events. Most of its tenants are small and medium sized high-tech firms or providers of high-tech services. Only a small number are start-ups in need of incubation, business counseling and mentorship, services provided by the TEC management team. Few of those firms are university spin-offs. TEC is certainly successful as a unit of Calgary Technologies Inc., the local technology economic development organization. In 2000, it was estimated that over 350 companies (many with three or four employees) had benefited from its programs, most as non-resident members. However, its role as an incubator is probably secondary to its status as an up-scale multi-tenant building, the center of activity of Calgary Technologies Inc. As noted by tenants during a series of interviews, it is a prestigious address, offering very nice modern facilities with an excellent telecommunication infrastructure, good networking opportunities, and help in getting access to local business services, mentors and financial resources.

Canadian experience with formal technology incubation is therefore mixed, at least until around the year 2000, with some award-winning incubators and others less successful. As noted in Chapter 3, other incubators have been created during the first years of the new millennium, in cities such as Ottawa (for example OBIC, a biotechnology incubator for early-stage research-intensive life-science and biotech firms; a branch of Innocentre, a not-for-profit organization dedicated to helping high-tech start-ups, which started in Montreal in 1987), Saskatoon (Industry Partnership Facility at NRC's Plant Biotechnology Institute), and Calgary (another branch of Innocentre). Innocentre aside, the most important characteristic of these incubators is their emphasis on public–private research partnerships, as the business aspects of incubation are often left to external organizations such as the Entrepreneurship Centre in Ottawa, local economic development organizations (networking, mentoring), or private service firms.

Table 5.10 Comparative overview of selected science and technology parks

Incubation mechanism assessment	Small cities		Medium sized cities		Big cities		Mexico City
	Querétaro–Bajío		Cuernavaca	Ensenada	Guadalajara	Monterrey	
Incubator/Science Park	PIEQ/San Fandila	CENIT	CEMIT, IETEC	IEBT	CUNITEC/Tech Park Belenes	ITESM entrepreneur's program	Siecyt-UNAM, 'Torre de Ingeniería'-UNAM, CETAF-UACH, CIEBT-IPN e INCUBASK-UAEM
Facility Background ● Type of mechanisms ● Sponsors, years established ● Tenant firms and employees ● Reasons for closing	Operating during 1992–1999. It had 3 graduated firms out of 25 tenants. Located in the Science Park of San Fandila, which did not take off.	Established in 1994 at Delta Industrial Park (Leon, Gto). In 1996 moved as an Industry–University Collaboration Program (VEN, Silao, Gto).	CEMIT founded in 1991 and IETEC (ITESM) established in 1994. Both were in operation until 1998.	Started operating in 1991 until 1998.	Created in 1992 at 'Los Belenes' Tech Park. There are 23 tenants, out of which 8 graduated. The Technology Park has 13 firms.	Entrepreneurship ITESM program started in 1985. There are 3 second generation incubators operating: San Pedro Garza García (2001), one ITESM incubator (2001), and UANL incubator (2002).	5 Incubators. The first 3 which were pioneers went bankrupt. IPN's and INCUBASK are operating. In INCUBASK, one out of 10 tenants graduated. The second generation have more types of firms.
Performance outcome ● Program sustainability, growth ● Firms survival,	Causes of failure are the absence of financial resources to support feasibility activities of	Now the incubation of tech firms is only one of its multi-entrepreneur functions of VEN.	No funding is available for costs compensation. The strong Cuernavaca	Causes of closing are due to lack of risk capital. Incubator is more a result of	There is no seed and risk capital available. The regional impact is low. But due to the sustainability	There is no risk and seed capital. Teaching entrepreneurship is an important activity of ITESM.	Lack of seed and risk capital are the main causes of first generation incubators closing.

growth, innov. ● Contribution to sponsor's mission ● Regional impacts	innovations as well as its commercialization. Good expectation on regional impact, but soon the Real State Development hosting a Research Center diminished its attraction to firms location in it. While regional impact was expected to be positive, some real estate acquisition issues hampered firms' relocation into the facility.	The firms' creation has very low profile considering the regional impacts.	research and development infrastructure has been an aftermath of the research centers located in Mexico City, due to its closeness.	technology push than a pull of the local demand, which is feeble for this kind of project.	of both incubator and tech park the example impact is high.	Then the second generation of incubators is an obligated output of this program.	The impact of incubators is marginal. It is more a push idea than a reality.
Management policies & their effectiveness ● Goals, structure, governance ● Financing and capitalization ● Operational policies ● Target markets	Querétaro State Government leads PIEQ's activities, jointly with CONACYT support. So this center has government leadership. CONACYT and NAFIN provided financial backing for 3 years.	CENIT was initially led by the municipality of Leon and by a research center. It came under university leadership when it moved to VEN at Guanajuato University. The main funding was provided by CONACYT, NAFIN and State government of Gto.	CEMIT has a large board of governors, which turns out to be an obstacle for flexible decision making. A research center (IIE) provided leadership in innovation and a federal government agency (NAFIN)	As other pioneer incubators the board of governors was large, but the leader institution is a Research Center (CICESE). Incubator set up was made possible by CICESE financial support. Additional risk capital came from	Guadalajara University is the leader institution and could have a flexible decision management.	Universities are the main agents settling incubation spaces (ITESM and UANL). There is a municipal local government incubator, the San Pedro Garza Garcia.	Universities (UNAM, IPN) are the main leaders in Mexico City incubators. However, they failed to make timely decisions aligned with firms' competitive environments to achieve business objectives.

Table 5.10 Continued

TBIM assessment element	Small cities		Medium sized cities		Big cities		Mexico City
	Querétaro–Bajío		Cuernavaca	Ensenada	Guadalajara	Monterrey	
Incubator/Science Park	PIEQ/San Fandila	CENIT	CEMIT, IETEC	IEBT	CUNITEC/Tech Park Belenes	ITESM entrepreneur's program	Siecyt-UNAM, 'Torre de Ingeniería'-UNAM, CETAF-UACH, CIEBT-IPN e INCUBASK-UAEM
		CONACYT, NAFIN and State government of Gto.	supplied funding. IETEC has been an ITESM local campus incubator.	Conacyt–Nafinsa fund.			
Services and their value added ● Shared services ● Networking support ● University-related services ● Tenant perceptions ● Other	Incubator help in sharing information, to create a positive tenants association to meet outside demands. Due to cash shortages often have problems covering incubator rent and common services.		Cuernavaca has a better research and development knowledge infrastructure and keeps a good knowledge networking.	Incubator has a good support from CICESE in knowledge networking.	Machine tool anchor firm helps the incubation of other firms. Incubator strong support for outside networking.	The second generation of incubators have lower risk tenants as most of them are not TBF. So the incubators service costs are mainly shared by tenants.	

Overall assessment ● A process view and performance challenges	The lack of seed funds and capital risk has a general effect on tenants' cash obligations. Potential tenants firms networking with the large research infrastructure is not fully utilized for technological innovation.	Incubator bankruptcy due to 1) lack of seed and risk capital; 2) rigid and slow decision making due to overloaded institutional governance; 3) strict breakeven for financial self sufficiency (IETEC).	The sustainability of the incubator and Tech Park is due to 1) flexible decision making, 2) ad-hoc tenants support in outside relationship.	Better chances of success due to a mixture of TBF tenants with traditional firms. Availability of public funds for spin-offs (AVANCE), as well as State Gov support.

5.3.3 Mexico's Science Parks and Incubators

Each of the four Mexican cases (six TIPs) included in this study has technology-based incubators, but only two (Querétaro–Bajío and Guadalajara) have science parks, one of which closed in 1995 (San Fandila, in Querétaro). Several of these incubators and science parks have encountered financial difficulties after some time in operation that have forced them to close. An overview of the incubators and science parks is provided in Table 5.10.

There have been two incubators in the Querétaro–Bajío region. The PIEQ enterprises incubator program, founded in 1992, was in operation in Querétaro until 1999. CENIT, the Technology Incubation Business Center, founded in the Bajío region's state of Guanajuato in 1996, moved from a research center in the city of Leon to become associated with an entrepreneurship and liaison program that was co-sponsored by the University of Guanajuato, in the city of Silao (INCU-VEN). There was also a science park, the San Fandila Technology Park, which was developed in the Querétaro region and became bankrupt by the end of 1990s.

Technology-based incubation centers are also found in the mid-sized cities included in our study, Cuernavaca and Ensenada. In Cuernavaca, the Technology Incubator Enterprises Center (CEMIT) was created in 1991 and the Incubator of Management and Technology Innovation Firms (IETEC) was created in 1994. In Ensenada, the Incubator of Technology Based Firms (IEBT) was founded in 1991. These three incubators operated with some success until 1998.

In the larger city of Guadalajara, the Industrial Complex houses the Technology Park 'Los Belenes', which belongs to Guadalajara University. The University Center for Technology Innovation (CUNITEC), was established in this park in 1992. In the city of Monterrey, the Entrepreneurship Program of the Monterrey Institute of Higher Education (ITESM) was founded in 1985, creating a culture of creativity and entrepreneurship. Other universities also launched entrepreneurial programs such as university courses which started at UANL in 1991. And in the early 2000s, several incubators were founded, including the San Pedro Garza Garcia business incubator in 2001, and ITESM's and UANL's incubators in 2002.

Four innovation zones can be observed in the metropolitan area of Mexico City. The first is in the southern part of the city around the Universidad Nacional Autonoma de Mexico (UNAM) where two incubators, Siecyt (Scientific and Technological Firms Incubator System) and the Engineering Tower, were founded respectively in 1991 and 1993 at UNAM. The second innovation zone, in the northern part of Mexico City, has ties to the National Polytechnic Institute and other research organizations such as the oil research institute (IMP). Here, a technology-based incubator center called INCUBASK began its operations in 1995 in Tecamac. It is managed by the State of Mexico University (UAEM). There is also the Ferrería High Technology Park project, which emerged in 2001 and is sponsored by the Federal District Government. The third zone is a cluster specializing in agricultural research, and is located towards the east in Texcoco, State of Mexico (see Chapter 4), where the Agriculture and Forestry Technology Firms Center (CETAF) was established at Chapingo's University in 1993. This center was closed soon thereafter due to managerial difficulties. There is a fourth innovation zone in Santa Fe, in the western part of Mexico City where the Iberoamericana university is located. This cluster is oriented towards modern and corporate financial services.

Table 5.11 lists other technology-based incubators in Mexico. The incubator of the University of Colima is focused on digital computer technologies, while that of the CIBNOR in La Paz BCS is focused on aquaculture, but the latter is stagnating and has not been hosting any firms since 2003.

Considering the dates when incubators were founded, three periods are evident in Mexico:

- The 'take-off' period, 1990–1994, characterized by high expectations, which saw the creation of several pioneering incubators, including in 1990 CEMIT at Cuernavaca City, in 1991 IEBT at Ensenada, and in 1992 alone, Siecyt in Mexico City, PIEQ in Querétaro and CUNITEC in Guadalajara. However, optimism fell quickly thereafter, and, in 2004, only the Guadalajara incubator was still in operation. Most incubators were shut down after 1995. They were affected by the economic crisis that began in December 1994 and also by a number of other problems including the lack of venture capital (both seed and growth capital), the rigidity of the decision making process, and the weak linkages they established with technology and knowledge related institutions. It was during this first period that the Mexican Association of Technology Parks and Entreprise Incubators (AMIEPAT) was created, which served as an umbrella organization established to represent and promote technology incubators in Mexico.[14]
- The 'transition' period (1995–2000), with limited incubation activity, is characterized by a change in emphasis from technology incubation only to general business incubation. During this period, incubators clearly decreased in number although they did not disappear altogether, as evidenced by a few isolated efforts such as the IPN's technology enterprises incubator center created in 1995, and INCUBASK, created in 1998.
- The 'entrepreneurial' period started in 2001 when incubators were created with the objective of founding and developing new enterprises, including technology-based firms and other types of firms, in all sectors

Table 5.11 Mexico: technology business incubators, 1990–2004

		Incubator name and status in 2004 (see note)	Year founded	Year closed	Lead institution		Firms			Science park
							Hosted	External	Graduated firms	
Region I	Querétaro	PIEQ (K)	1992	1999	State Gov (G)	G	25		3	Fandila (K)
	Bajío	CENIT (S)	1996		UGto (U)	RC–>U	4			—
Region II	Ensenada	IEBT (K)	1991	1998	CICESE	RC	8			
		SIETAI-Mexicali (C)	1992	1994	UBC	U				
	Cuernavaca	CEMIT (K)	1990	1998	IIE	RC				TPM (p)
		IETEC-ITESM (K)	1994	1998	ITESM	U	5	6		Tech Park Morelos
Region III	Monterrey	Entrepreneurship, ITESM (+)	2001				8		2	—
		Inc. System (31 campus) (N, +)	2002		ITESM	U				
		Inc UANL (+)	2002		UANL	U	8			
	Guadalajara	San Pedro Garza García (N, +)	2002			GM	32		6	
		CUNITEC (+)	1993		UdG	U	34		8	Los Belenes*
Region IV	Mexico City	SIECYT-UNAM (K)	1991	1998	UNAM	U				Ferrerias (p)
		Cetaf-UACH (P)	1992	1993	UACh	U				
		Torre de Ing-UNAM (S)	1993	1999	II	RC		65		
		CIEBT-IPN (+)	1995		IPN	U	17			
		INCUBASK UAEM-Tecamac (+)	1998			GI	8		1	
	Other	IEBT-Colima (+)	1994		U Colima	U	4			
		IEBT-Yucatan (C)	1992	1994	State Gov	G	1			
		IEBT-CIBNOR (+)	1994		CIBNOR	RC	0			
		CIDET-Toluca (C)	1993	1998		Private	6			

Total U = 10 RC = 4 G = 4 Private = 1	160	71	20	4

Notes:
Status of the incubation mechanisms in 2004:
Planned (P); Closed (C); Bankrupt (K); Surviving (S); Operating (+); (N) indicates a non-technology incubator

of the economy. During this period, the San Pedro Garza García incubator (2002) and the UANL incubator (2002) were established in Monterrey metropolitan area. Also ITESM University with its main campus in Monterrey established 31 incubation facilities on several of its campuses around the country. In addition, the Ferrería High Technology Park was created in Mexico City during this period. In August 2003, AMIEPAT was replaced by the newly created Mexican Association of Management Incubation (AMIRE). This new professional association seeks to generate employment and economic development by stimulating business creation, growth and consolidation. It does so by emphasizing sectoral and regional competitiveness, by supporting initiatives for the creation and successful operation of high value added enterprises, technology-based firms, and by establishing knowledge management networks, science and technology parks, private capital investment firms, industry clusters and strategic alliances. And in 2004, the Federal Government's Ministry of Economics launched a program for incubating small and medium sized enterprises (SMEs). By creating incubators associated with selected municipal governments, this program aims to install at least one incubator in every state. Several federal authorities will be working together on this project, including the Public Education Department and National Financiera (CRECE-SEP-ANUIES-NAFIN). The goal is to serve 4000 entrepreneurs during 18 months, thus creating 2800 SMEs.

Incubated firms

The 20 incubators listed in Table 5.11 have had different degrees of success and widely varying results. They host from one to 34 tenant firms. Some incubators are working at full capacity with about 20 firms, but occupancy in most of the others has usually been limited to about half of this. Reported graduation is about 20 firms during the five-year period of operation, but this may be overestimated because some firms moved out long after reaching a successful operating capacity. It is estimated that over the last decade an average of 10 incubation facilities have been in operation at any given time (this is by adding new facilities that were established and subtracting the ones that closed).

Causes of incubator failures

The failure of the Mexican incubation programs during the 1990s can be attributed to a number of factors. The shortage of seed capital and risk investment required for both the development of the incubating firms and the operation of the incubators themselves is a direct cause of many incubation failures. Other causes already mentioned above added to the challenges faced by these programs. They include rigid management structures that stifle timely and flexible decision making processes, low levels of networking with knowledge institutions, and short-term government policies resulting in the drying up of financial support.

As of 2004, the regional impact of incubators has been marginal at best, resulting in low industrial visibility. The impact is more visible among TBFs located in regions that have an entrepreneurial environment, probably due to the fact that local industries' needs influence the technology focus of incubators.

Financial backing

Until 2003, the main source of funding for technology firms was the FIDETEC, a joint partnership between CONACYT and Nacional Financiera (NAFIN), which can grant financial aid under specific circumstances. In 2002, Conacyt initiated an innovative program supporting new business creation by funding technological and scientific developments (AVANCE).[15]

Sectoral and mixed funding programs have been created to provide financial support to the science and technology research effort. The sector-related funds, built up since 2002, are a joint creation of CONACYT, and federal public management branch offices, which provide funding for science and technology research in their sectors of competence. Mixed funds, created in 2001, are the result of an agreement between CONACYT and each state, through which they jointly provide financial backing to appropriate science and technology research projects. The research supported by both programs falls into the categories of information technology, electronics and telecommunication, health (diagnostic systems, equipment and drugs), land and cattle, fishing and foodstuffs, advanced materials, sustained development and environment, energy, design and manufacture, dwellings and building, poverty and social development, and health care.

It is perhaps not a coincidence that mixed and sector funds, as well as the AVANCE program, were created during the 'entrepreneurial' period of incubator creation, described above. By providing financial support, these initiatives contributed indirectly to the development of the incubation centers.

5.3.4 Synthesis and Learning

As shown by the preceding review of selected TIPs, the rate at which new science parks and incubator facilities are created seems to have reached a plateau in NAFTA countries, a sign of the relative maturity of the industry (Chapters 2–4). At the same time, however, there are indications that the technology innovation pole model, which uses parks and technology incubators as enablers of development, continues to be a favored policy instrument for technology-based economic development (TBED) in many North American regions.

Our comparative analysis of the 12 NAFTA cases shows that there are significant differences between regions in terms of incubation mechanisms. The Canadian and US cases offer examples of successful science parks while Mexico's experience with science parks is very limited. The US cases also offer examples of mostly successful incubator models, while the Canadian and the Mexican cases contain a mix of successes and failures. For Canada, this is more a representation of the TIPs selected for analysis in this book than of the Canadian technology incubation industry, as there are other successful incubators in the country, which are not included in this sample.[16]

As shown in the examples discussed above, the major R&D institutions and/or research universities associated with most incubators play a significant role in securing new knowledge and resources, and also provide champions or sponsors within the system. The surrounding community, including private industry and the public sector (federal, state, and/or local levels) provide funds, volunteers, additional services and other forms of in-kind support. The incubators host qualified entrepreneurs and new start-up firms as tenants, and provide a nurturing environment, networking opportunities and value-added services that help them to develop and grow. The resulting newly established firms and experienced entrepreneurs (incubator graduates) then move to an associated science park or somewhere else in the community to continue their technology-based activities.

In terms of governance, most US technology parks are led by a board whose members are key stakeholders and provide policy guidance. Typically, the park's management team is headed by a director to whom the incubator manager reports, as is the case in the four parks studied. However, in a number of other US facilities, incubators operate independently of the parks. In Canada, the boards of all the science parks surveyed include representatives from universities, research centers, various levels of government, and the private sector, and the parks themselves tend to be managed by independent non-profit organizations. The Canadian technology incubators presented in this study operate independently of parks and universities, although this is not the case for all technology incubators in Canada.[17] In Mexico, incubator management boards tend to be based on partnerships with a university, a government (at the federal, state or municipal level),[18] and some private institutes. Universities have been the dominant institution in half of the incubators, two being directed by state government, one by the local government, and another which used to be promoted by a foreign region, the Basque province of Spain (INCUBASK) (Table 5.11). In all cases, key to park success is the provision of non-monetary incentives such as the availability of value-added services (often from independent private firms) and R&D

(geographic and sectoral concentration of R&D activities, proximity to a university or a large research laboratory, presence of large anchor organizations). Monetary incentives such as below-market rental/leasing prices are generally avoided as they attract organizations that may not be otherwise viable.

In summary, by relying on incubation tools in developing successful TIPs, these regions have attempted to address the existing gaps in innovation support activities in their communities, providing an environment that included all the necessary ingredients described earlier (Chapter 1). In this milieu, science parks and incubators served as rallying points and networking hubs within the respective regional environment.

5.4 PERSPECTIVES ON INNOVATION AND TIP IN NAFTA

This section examines the perspectives of the technology innovation poles, describes their trends in each NAFTA country, and concludes by suggesting a scenario of cooperation for innovation.

5.4.1 US: Perspectives on TIP, Science Parks and Incubation in NAFTA

US regions are using complex collaborative and competitive strategies to shift away from their dependence on 'old economy' firms and to promote the growth of 'new economy' firms that rely on knowledge and innovation to sustain and improve competitiveness.

As described earlier (section 5.2), each of the US regions studied has its own unique characteristics, but most of them had somewhat similar socio-economic profiles prior to their development as TIPs. Pre-existing characteristics include the presence of a skilled workforce, well-developed institutions of higher education, an active private industry, a supportive government sector, and indicators of a good quality of life. Given these preconditions and the assessment that some economic activities had to be restructured, regional leaders opted for concerted action. Their efforts were often undertaken in partnership with academia, industry and government – and the extent of involvement of each of these entities differed from case to case depending on the presence in their midst of committed champions. The leading role of the academic sector is evident in the New York Capital Region, the New River Valley Region and the Madison Metropolitan areas. In contrast, in the Chicago PMSA, the local city government of Evanston was an equal partner with the local university.[19] These partnerships led to the development of a technology infrastructure in which inter-disciplinary research centers, science parks and incubators were key elements.

As part of their regional economic development initiatives, these regions have built upon their respective historic strengths and targeted certain technology sectors. The four cases studied show slightly different trends. The New York Capital Region and Virginia's New River Valley Region have emphasized diverse technology sectors, whereas the Chicago and Madison Metropolitan Areas have focused on biotechnology and information technology. At the same time, however, the university-led park and incubator facilities in all four areas have followed the national trend in nurturing a large number of information technology- and biotechnology-related firms. After years of concerted efforts, these regions have developed a comprehensive infrastructure consisting of knowledge institutions (for example, entrepreneurial universities/colleges), major R&D centers and research laboratories, a significant number of high-tech firms, successful science parks and incubators, entrepreneurs and their new technology-based start-ups, and good sources of risk capital.

A series of benchmarking characteristics for successful US TIPs has been developed by combining into key strategic dimensions certain aspects of innovation research and regional science (Cooke et al., 1991). These dimensions include:

a. national and international visibility of the region, allowing it to compete successfully for technology-based growth
b. connectivity within the region and between the region and the outside world
c. a plentiful, well-educated and productive workforce coming from the region's universities
d. a positive entrepreneurial climate encouraged by a supportive state and local government, engaging private sector and university leadership, and the availability of capital, R&D and innovation, intellectual property and entrepreneurial talent
e. ongoing diversification of the regional economy by emphasizing cultivation of technology-driven companies.[20]

The inability of certain developed regions to stimulate a stagnating economy may be due in part to their slow population growth, the result of lack of economic opportunity and/or a harsh climate. However, in this study, these factors were found to be largely balanced by the generally superior work habits and discipline of the labor force, as reported by several of the entrepreneurs interviewed in the New York Capital Area and the Chicago PMSA. It may be that, while promoting a technology-based entrepreneurial climate, the business culture in these two regions may have deterred some firms from locating there because of a perception that non-technology types would be difficult to recruit. These regions may also have benefited from their good infrastructure and availability of venture capital, advantages that may not have been present in less-developed regions. The presence of major research universities with land-grant missions played a significant role in the transformation of the Madison Metropolitan Area and the New River Valley Region. These cases suggest that regions with a traditional economic base and a favorable quality of life can acquire a vibrant technology-base if they have a committed higher education sector with adequate public and private support. Besides broad government support for the development of infrastructure, successful US regions have also benefited from various economic regeneration and development programs at the federal, state and regional levels (see section 2.1). These programs provide grant money to stimulate regionally-led initiatives such as incubators and small business development centers that help to promote investment and stimulate an entrepreneurial climate.

5.4.2 Canada: Perspectives on TIP, Science Parks and Incubation in NAFTA

The successful development of the Canadian TIPs is due to a number of key factors, including:

- their solid research base (government research laboratories, private research centers, universities, research hospitals);
- the presence of a champion, a proactive development organization or consortium (public or not-for-profit) that facilitates linkages and networking;
- support for the development of an entrepreneurial culture, and encouragement of government–industry–university cooperation;
- the presence of one or more large research/manufacturing private firms (a large 'anchor organization');
- the presence of at least one good research university (source of highly trained manpower and of good research);
- a good quality of life;
- and good physical communications with the United States and the rest of the world.

Public policy measures (generally at the national/federal level) are also of prime importance. They are designed to encourage university–industry research cooperation, to facilitate access to technology by SMEs, to increase the supply of venture capital, to develop, attract and retain top managerial talent (especially technology leaders with a vision), and to develop, attract and retain top expertise in technology sales and marketing.

A number of regions started out without some of these desirable attributes, and regions where they were present were more likely to be successful at developing the critical mass of high-tech activity necessary to be visible and attractive to new technology businesses. In some regions, desirable attributes that had been absent (for example a large anchor organization, venture capital, business services) developed over time, emerging out of other attributes or in parallel with the development of the region.

Science parks have proved to be beneficial mechanisms for innovation in each region, either as the focal point for R&D and high-tech development (Calgary, Quebec, Saskatoon), or by encouraging the development of a specific sector (Ottawa). In all cases, success depended on a number of factors, including the reliable cooperation of all levels of government, industry and universities; the presence of active research centers and laboratories in or near the park, in such a way as to create a critical mass in research; the leadership of a champion (often a public–private–university consortium); and a value proposition based on the availability of services – such as the intensity of knowledge activities, networking opportunities, proximity to research laboratories and to a university – rather than mere monetary incentives.

None of the research universities in the Canadian regions surveyed had a technology incubator, although, in the early 2000s, interest for such university incubators was growing as they were increasingly seen as efficient mechanisms to reinforce university–industry linkages and facilitate the commercialization of university technology. Rather, two of the three most successful incubators in the regions surveyed were located in large public research laboratories and served as efficient technology transfer facilities, and the third was located in a luxurious multi-tenant building in a science park. The first two incubators do not hesitate to charge above market rents and fees for the exceptional access they provide to researchers, technology, technical services and secure space; the third incubator charges above market rent for its high quality secure space and colocation with other high-tech firms and services. In most of these cases, external organizations serving the region and operating as 'virtual incubators' (for example, Ottawa's Entrepreneurship Centre) offer small business support and advice, opportunities for technology and business networking, mentorship, help in accessing angel investors, and venture capitalists. As a result, it is not absolutely necessary to offer such services and support in the parks and incubators themselves, though having such dedicated facilities that are well conceived and professionally managed have clearly shown positive results in a number of our cases studied here.

The results of our study of TIP development in Canada can be summarized in terms of several recommendations:

a. Keep on focusing on R&D to attract high-tech firms and high-tech development;
b. Have a champion to promote government–university–industry partnerships, and facilitate networking between all persons and organizations involved;
c. Attract some large 'anchor organizations' to the region; and
d. Keep the region competitive in terms of quality of life and physical communications (air links in particular).

A similar summary could be used for the development of science parks and technology incubators, with one additional recommendation:

e. Have a value proposition based on research, networking, access to knowledge and services rather than on monetary incentives (i.e. avoid low rent or low fees).

Successful TIP development also benefits from policies at the national/federal level designed to encourage venture capital investment and to develop, attract and retain top talent in high-tech firm management and in technology sales and marketing.

5.4.3 Mexico: Perspectives on Innovation and TIPs in NAFTA

Mexico's science and technology infrastructure is quite different from that in the US and Canada in both quantitative and qualitative terms. In quantitative terms, Mexico needs:

- To develop the country's research base,
- To increase significantly its research activities,
- To promote industrial research among large and small firms,
- To train more technological and scientific human resources.

Qualitative improvements can arise from increasing the efficiency of Mexico's national innovation system, improving the way in which universities, the private sector, and public policies interact to produce economically meaningful innovation.

The real question is how to do it, how can the country begin to address these issues to be ready to capture opportunities that may arise?

Mexico needs to improve the quality of its research institutions (research centers and universities). Most of the science and technology indicators show that Mexican innovation performance is significantly below that of the other NAFTA countries.

In order to catch up with the pace of innovations in the US and Canada, Mexico also needs to make substantial policy improvements – NAFTA is not enough. Rather than importing a 'foreign model' for science and technology systems, Mexico must develop alternative paths, whose main aspects can be summarized as follows:

- Priorities: Mexico has focused first and foremost on building capabilities and reaching a 'critical mass' for research and development. Yet, even as it follows this policy, certain qualitative changes must take place. These include giving higher priority to the indigenous technology transfer and commercialization of technologies, enhancing local absorptive capacity through quality manpower training and inter-nationalization of all types of knowledge relationships.
- Selectivity: Mexico must focus on certain regions and on a limited number of sectors. The emphasis on technology transfers should reflect the country's relatively small scientific and technological infrastructure, and Mexico can also improve and accelerate its local capacities for innovation by working out an aggressive policy for multiple regional and institutional networking within NAFTA countries.
- Orientation: The possibility of gaining access to and attract NAFTA funds could create incentives for SMEs to participate in the flow of knowledge and the commercialization of technology. In this respect, policies should encourage spin-offs and labor mobility among researchers and qualified technology personnel. They should also promote collaboration between universities, firms and the three levels of government (national, state and local), emphasizing awareness about SMEs as key in generating a sustainable flow of innovation and new products and technologies.
- Networking: It is rarely clear who initiates the development of a TIP, whether the government (local, state and/or national), a research university, a research center or a firm. In Mexico, research centers have often been the first to actively pursue innovation, but they have failed to enlist the participation of other institutions in the development and innovation of technology. Clearly, active pursuit of innovation by some institutions must be followed by a process that will facilitate the development of knowledge linkages with other institutions. In order to encourage networking among innovators, some public funds are needed that will create incentives for cooperation or even partnership between government entities, universities and private firms.

Where can new forces of change come from?

The results of Mexico's 10-year membership in NAFTA (1994–2003) have been summarized as follows: 'Mexico's global exports would have been about

25 per cent lower without NAFTA, and foreign direct investment (FDI) would have been about 40 per cent less without NAFTA. Also, the amount of time required for Mexican manufacturers to adopt US technological innovations was cut in half; and Mexico's $5,920 per capita income in 2002 would have been 4–5 per cent lower'[21] (Lederman et al, 2005b). Clearly, 'trade liberalization and NAFTA are helpful but they are not enough to help Mexico catch-up to the levels of innovation and the pace of technological progress observed in its North American partners, especially the United States.' (Lederman et al., 2005b).

Mexico is likely to be affected by US free trade agreements with other Latin American countries, including the FTAA, the Central America–US Free Trade Agreement (CAFTA), and an already existing agreement between the US and Chile.

It is expected that Mexico's economic position will suffer as a result of these new agreements. Consequently, it is in the country's interest to increase its participation in NAFTA science and technology networking, especially that having to do with the commercialization of technologies. More specifically, Mexico should expedite the negotiations for co-financing of research exchange programs with its NAFTA partners.

Mexico must also improve its international position in science and technology, as it might be able to capture new opportunities if it played a 'pivotal role' in NAFTA, and could set itself ahead of other Latin American countries to operate as a 'hinge' in the FTAA.

Firms: Mexico needs to increase the participation of firms in innovation. Until now, Mexican firms have played a minor role in the process of innovation. Section 5.1 noted that the share of R&D financed and performed by industry is 29.8 per cent and 30.3 per cent respectively in Mexico, 63.1 per cent and 68.9 per cent in the US, and 44.3 per cent and 53.7 per cent in Canada. In order to address this problem, it is useful to examine some Mexican success stories, including the development of an incubator in a large firm in the early 1990s (Resistol), and the growth of an electronics and computing cluster following the location of some multinational firm facilities (including IBM) in Guadalajara in the 1950s. These case stories show that participation in innovation can involve both large firms and SMEs, and also that it can take place on an international, global and/or national level depending on a variety of factors.

Bridging institutions: Mexico has developed a number of bridging or linking institutions for technology transfer in which adequate public incentives must be provided to promote standardization of industrial products and processes; to support regional venture capital in diminishing the risks of innovation; to facilitate technology transfers; to stimulate cooperation between research centers and universities, possibly by creating research consortia; to take part in large scale scientific projects; and to appropriate knowledge. In doing so, these institutions should build on existing accumulated capabilities and aim to develop them in specific sectors and regions of the country. In this way, the reported TIPs could shape new paths in dynamic innovation activity.

Additionally, Mexico's relatively low innovation capabilities in relation to NAFTA could be improved by taking advantage of its privileged location with respect to its more industrially accomplished partners and existing communication channels. This can be done by facilitating access to technology, and enhancing Mexican capabilities as a receptor of specific technologies. Ultimately, Mexico's goal should be to design and implement a strategy based on the development over the next decades of a small number of TIPs in high and medium technology sectors.

5.4.4 Potential for Cooperation between TIPs in NAFTA

The key factors that enhance the creation and diffusion of innovations seem to be the amount of knowledge available and produced locally, the intensity of networks, an institutional environment that encourages entrepreneurship, the availability of venture capital and of business and technical services, and, as additional enablers, flexible and timely support mechanisms including incubators and science parks. The presence of these factors facilitates the development of a pattern of cooperation and networking that is conducive to innovation. Cooperation and networking does take place within successful TIPs. It also occurs nationally and internationally between TIPs. The context of NAFTA and the broad collaborative environment it offers member countries provides an opportunity for enhanced cooperation between knowledge regions and TBFs, to benefit from each country's, each region's comparative advantage, whether a high level of expertise in a specialized niche, a favorable R&D tax treatment, a low cost structure, direct access to customers, or other factors enhancing some aspects of the creativity/innovation/commercialization process.

For the US, with a relatively high cost of high-tech human resources, it may be attractive to reallocate some R&D activities to other less costly NAFTA regions. Taking advantage of less expensive, well trained scientific human resources may be a real benefit. In the US–Mexican context, the often-controversial – yet mutually beneficial – maquiladora program has laid the foundations for a tangible collaborative business environment. As a next step, however, it is necessary to move beyond the maquiladora initiative and to find ways to engage Mexico's burgeoning knowledge and innovation sector in a mutually beneficial way. Within the US, efforts are already underway to outsource US service jobs to other Asian and/or Latin American markets (in the event of the envisaged expansion of NAFTA). In this context, it may be useful for the US and Mexican administrations to reach an understanding and follow it with concerted policy actions to replace the maquiladora program with its emphasis on low labor costs by a science and technology program designed to take advantage of Mexico's well trained, low-cost scientists and engineers.

The post-September 11, 2001 emphasis in the US on security also has an impact on national economic priorities and on the mobility of international scientists, both with significant implications for the US civilian sector. Relocating some research activities to Canada or to Mexico or partnering with research centers and universities located in those two countries would reduce the impact of these changes, facilitating access to foreign scientists and expertise by US companies. The potential use of Mexico's emerging TIPs (discussed in this book) as launching spaces for US and Canadian TBFs and R&D organizations may open up new doors for collaboration based on science and technology.

Canada's current objectives in science and technology are to keep on improving its national system of innovations by continuing to encourage private sector spending on R&D, to increase its share of GNP and narrow the gap with its most active partners; to provide incentives and support for commercialization with a special focus on SMEs; to further encourage the growth of venture capital investments through favorable tax treatment of high-tech investment; and to continue to build research partnerships between universities, government laboratories, and industry, a trend that is encouraged by a renewed interest for incubators at universities. One major concern in the country in this post-September 11, 2001 era will be to keep the US border open to Canadian goods and services, including high-tech goods. Historically, the Canadian high-tech sector has had very strong ties to the US. Many Canadian TIPs have been using publicly financed trade missions or other high-level visits to develop special partnerships with US TIPs in similar or complementary industries. Many Canadian high-tech companies have sales offices and/or part of their operations in the US. Continuing to build linkages between Canadian and US TIPs and encouraging US firms to relocate some of their R&D activities to take advantage of Canada's generous R&D tax credits is one way to keep the US–Canada border open.

Until now, Mexico's less developed market for high-tech goods and services has not been very attractive to Canadian firms. This suggests that Canadian collaboration with Mexico is more likely to arise from Mexican initiatives than from Canadian ones, although it could result from bi- or trilateral cooperation involving the US as well. Some Canadian TIPs, particularly in the areas of telecommunications, energy, pharmaceuticals, biotechnology and ag-biotech, would certainly gain from more cooperation with Mexican TIPs in similar domains insofar as they would be encouraged to adapt technologies and innovations to a different environment. The foundations already exist for such cooperative endeavors but the number of exchanges, partnerships and collaborative agreements could be increased. A few of the many examples of technological complementarities that can be exploited include agri-biotechnology between Saskatoon and Bajío, information technology between Ottawa, Guadalajara and Chicago-Evanston, and pharmaceuticals in the New York Capital Region, Montreal and possibly Mexico City. The Ensenada TIP already has ties to US high technology on oceanography facilities located in La Jolla, California.

Mexico needs to catch up with its NAFTA partners by generating opportunities to participate in large NAFTA science and technology projects that would enable it to develop its scientific and technological capabilities. Many questions remain about the way in which Mexico will best be able to establish cooperative innovation within NAFTA. Yet the advantages of working cooperatively within NAFTA clearly outweigh those of working alone. A NAFTA oriented supra-national outlook can promote the development of regional innovative clusters that attract qualified labor and foreign direct investment. As noted by OECD, profiting from globalization on a local or national level 'requires a regional approach to industry–science partnerships, since the nature of the international linkages to be developed depends on the characteristics of the innovative cluster' (OECD, 2004c). In this scenario, Mexico would hope to develop multiple cooperative agreements within NAFTA to provide its less fortunate regions with venture capital, entrepreneurship programs and technology incubation facilities.

NOTES

1. PPP dollars: US dollars estimated at their purchasing power parities, OECD.
2. Tertiary-level education above secondary school.
3. Ratio of US and Canada GERDs to Mexico's.
4. In 2003, in Ottawa there were 1043 high-tech firms (excluding engineering and business services firms), while in Silicon Valley there were 25 787, including professional and innovation service firms (Brouard et al., 2004, Table 8).
5. For details see Bradshaw and Blakely (1999).
6. For details see Technology Transfer and Commercialization: Their Role in Economic Development, 2003, EDA (http://www.eda.gov/pdf/eda_ttc.pdf).
7. Innovation Place (www.innovationplace.com) in Saskatoon and the Quebec Biotechnology Innovation Centre (www.cqib.org, in Laval, close to Montreal) have both won international science park of the year and incubator of the year awards.
8. When analyzing university–industry linkages, it is desirable to separate research linkages from the offer of standard services, often provided through consultancy, which do not necessarily imply R&D activities.
9. This also applies to other Mexican cities not included in this study. For example, CIBNOR in La Paz BCS, ININ Toluca, have been active research centers promoting technology incubators. Some firms, such as Resistol (now Grupo Girsa) in Toluca, were also active at the beginning of the 1990s.

10. The information was collected in multiple surveys and personal interviews conducted during 1998–2003.
11. http://www.parctechno.qc.ca/english/histo.htm (URL in March 2004).
12. This park was created in 1992 as a University of Ottawa initiative backed by the Province of Ontario and the City of Ottawa. It failed quickly because of lack of financial support and is now owned by the Ontario Development Corporation and has been managed since 1994 by the Ottawa Life Science Council (OLSC), a public/private partnership representing local life-science interests including local hospitals and the University of Ottawa.
13. Such as the Quebec Biotechnology Innovation Centre (www.cqib.org, in Laval, close to Montreal).
14. AMIEPAT's activities are: a) creating and developing technology-based enterprise incubators, research and technology parks, and other supporting tools for technology-based enterprises; b) Establishing technology-based firms in the academic sector, technology research and development, and industries; c) Working to favor commercialization, transference and development of technologies, being competitive in the relevant industry, and promoting new technological development and an entrepreneurial culture; d) Increasing social acceptance of the role that technology-based firms play in the country's economic development (<www.cfyge-prifepei.ipn.mx>, consulted on January 5, 2004).
15. According to www.conacyt.mx/dat/avance, on 26 November, 2003.
16. Such as QBIC, the Québec Biotechnology Innovation Centre (www.cqib.org/indexAnglais.htm), a prize-winning technology incubator in Laval, next to Montreal, operating in a park with solid links to several universities.
17. QBIC (see note 16), is an excellent example.
18. CONACYT (Federal government) and Querétaro state government officials are on the board of PIEQ (Table 5.11).
19. More recently the City of Evanston has withdrawn its role as one of the two financial backers while continuing to be a regional stakeholder in the project.
20. This dimension is derived from the four US case studies and does not focus specifically on the more prominent and extensively studied regions such as Silicon Valley in California and Route 128 in Boston, Massachusetts.
21. In US dollars rather than in PPP dollars (purchasing power parity) as reported in section 5.1 and Table 5.1.

References

Aghion, P. and P. Howitt (1997), *Endogenous Growth Theory,* Cambridge, Mass., MIT Press.

Alic, John (2001), 'Post industrial technology policy', *Research Policy*, **30**, 873–89.

Allen, D. and E. Bazan (1990), 'Value-added contribution of Pennsylvania's Business Incubators to Tenant Firms and Local Economies', Report prepared for Pennsylvania Department of Commerce, Pennsylvania State University, PA.

Almeida, P. and B. Kogut (1999), 'Localization of knowledge and the mobility of engineers in regional networks', *Management Science*, **45**(7), 905–18.

Archibugi, Daniele, Jeremy Howells and Jonathan Michie (1999), *Innovation Policy in a Global Economy*, Cambridge, UK: Cambridge University Press.

Audretsch, D. B. and M. P. Feldman (1996), 'R&D Spillovers and the Geography of Innovation and Production', *American Economic Review*, **86**(13), 630–40.

AURP (2003), 'University research park profile', Rockville, MD: Association Research Park Inc., January.

Barker, M. (1996), 'Modalities of U-I Cooperation in the APEC Region, Country Report for Canada', Report prepared for the Association of Universities and Colleges in Canada, August.

Berglund, Dan (1998), 'SSTI weekly digest', Columbus, Ohio: State Science and Technology Institute, website accessed February 6, http://www.ssti.org.

Berglund, Dan and Chris Coburn (1995), *Partnerships: A Compendium of State and Federal Cooperative Technology Programs*, Columbus, OH, Battelle Labs.

Bergman, Edward, David Charles and Pim den Hertog (2001), 'In pursuit of innovative clusters', in Bergman et al. (eds), *Innovative Clusters. Drivers of National Innovation Systems*, Paris: OECD, pp. 405–19.

Betz, Frederick (1994), 'Basic research and technology transfer', *International Journal of Technology Management,* Special Issue on Technological Responses to Increasing Competition, **9**, (5-6-7), 784–96.

Biggart, N. and M. Guillen (1999), 'Developing difference: social organization and the rise of the auto industries of South Korea, Taiwan, Spain, and Argentina', *American Sociological Review*, **64**, 722–47.

Bradshaw, Ted and Edward Blakely (1999), 'What are the third-wave state economic development efforts? From incentives to industrial policy', *Economic Development Quarterly*, **13**(3), 229–44.

Bresnahan, Timothy and Alfonso Gambardella (2004), 'Old economy inputs for new economy outcomes: what have we learned?', in Timothy Bresnahan and Alfonso Gambardella (eds), *Building High-Tech Clusters: Silicon Valley and Beyond*, Cambridge: Cambridge University Press, pp. 331–58.

Brouard, F., T. Chamberlin, J. Doutriaux and J. de la Mothe (2004), 'Firm Demographics in Silicon Valley North', in L.V. Shavinina (ed.), *Silicon Valley North, A High-tech Cluster of Innovation and Entrepreneurship*, Amsterdam, Oxford: Elsevier, pp. 57–84.

Callon, M. (1992), 'Dynamics of techno-economic networks' in R. Coombs, P. Saviotti and V. Walsh (eds), *Technological Change and Company Strategies: Economic and Sociological Perspectives*, London: Harcourt Brace Jovanovich, pp. 72–102.

Canadian Encyclopedia The (1988), Edmonton, Alberta, Canada: Hartig Publishers Ltd, second edition.

Canadian Newspaper Services International (1997), The Blue Book of Canadian Businesses, Ontario: Canadian Business Resource.

Casalet, M. (2000), 'The institutional matrix and its main functional activities supporting innovation', in M. Cimoli, *Developing Innovation Systems. Mexico in a Global Context*, London and New York: continuum.

Casas, Rosalba (1985), *El Estado y la politica de la ciencia en México*, ILS UNAM, México.

Casas, Rosalba, Rebeca de Gortari and Ma. Josefa Santos (2000), 'The building of knowledge spaces in Mexico: A regional approach to networking', *Research Policy*, **29**(2), 225–41.

Cassiolato, J.E. and Helena Lastres (1999), *Globalizaçao & Inovaçao Localizada. Experiências de Sistemas Locais no Mercosul*, IBICT-MCT, Brasilia, Brazil.

Castells, Manuel (1999), *The Information Era: Economy, Society and Culture. Vol. 1: The rise of the Network Society*, Oxford and Malden, MA: Blackwell Publishers.

Castells, Manuel and Peter Hall (1994), *Technopoles: Mines and Foundries of Information Economy Technopoles of the World: The Making of the 21st Century Industrial Complexes*, London: Routledge.

Chandler, Jr., A.D. (1997), *The Visible Hand*, Cambridge, Massachusetts: Harvard University Press.

Chrisman, James J. (1994), *Economic Benefits Provided to the Province of Alberta by the Faculty of the University of Calgary*, report prepared for Langford C.H., VP (Research), University of Calgary, September.

Cimoli, M. (2000), *Developing Innovation Systems. Mexico in a Global Context*, Science, Technology and International Political Economy Series, edited by John de la Motte, London and New York: Continuum.

Conacyt (2004), 'Informe general del estado de la ciencia y la tecnología en México', October, México, 274pp, www. conacyt.mx.

Cooke, Philip (2004), 'Regional innovation systems: an evolutionary approach', in Philip Cooke, Martin Heidenreich and Hans-Joachim Braczyk (eds), *Regional Innovation Systems: The Role of Governance in a Globalized World*, second edition, London, New York: Routledge.

Cooke, Philip, Mikel Gomez Uranga and Goio Etxebarria (1991), 'Regional Innovation Systems: Institutional and Organizational Dimensions', *Research Policy*, **26**, 475–91.

Cooke, Philip, Mikel Gomez Uranga and Goio Etxebarria (1998), 'Regional systems of innovation: an evolutionary perspective', *Environmental and Planning A*, **30**, 1563–84.

Cooke, Philip, Patries Berkholt and Franz Thodtling (2000), *The Governance of Innovation in Europe: Regional Perspectives on Global Competitiveness*, London: Pinter.

Corona, Leonel (1994), 'Hacia la consolidación de las empresas innovadoras', *Economía Informa*, Facultad de Economía, Universidad Nacional Autónoma de México, October, 232, 6–13.

Corona, Leonel (1997), *Cien Empresas Innovadoras en México*, MA Porrúa (ed), México.

Corona, Leonel (coord.) (1999), *Innovación Tecnológica y Desarrollo Regional*, Pachuca, Mexico: Universidad Autónoma del Estado de Hidalgo.

Corona, Leonel (2001), *Innovación y Región. Empresas Innovadoras en los Corredores Industriales de Querétaro y Bajío*, Querétaro, México: UAQ.

Corona, Leonel (2002), *Teorías Económicas de la Innovación Tecnológica*, México: CIECAS-IPN.

Covarrubias-Gaitán, Francisco (2000), 'Prospectivas de la urbanización en la Ciudad de México', *El Mercado de Valores*, Nacional Financiera, México, 4/Abril, pp. 3–19.

CREDEQ (2000), 'Répertoire des enterprises ayant bénéficié des services du CREDEQ', Janvier, Quebec, Qc, Canada.

Croft, W.D. (1997), 'Calgary, Canada, a model Technopole', presentation made at Technopolis 97, Ottawa, 10 September.

Davelaar, E. J. (1991), *Regional Economic Analysis of Innovation and Incubation*, Avebury: Gower.

David, P.A. (1975), *Technical Choice, Innovation, and Economic Growth*, New York: Cambridge University Press.

DeVol, Ross (1999), *America's High-Tech Economy: Growth, Development, and Risks for Metropolitan Areas*, California: Milken Institute.

Donnelly, Brian (1996), *US Research Parks and their Role in Technological Entrepreneurship*, paper presented at the *INFORMS Conference*, Atlanta,

Georgia, 6 November.
Dorfman, N. (1983), 'Route 128: The development of a regional high-tech economy', *Research Policy*, **12**, 296–316.
Doutriaux, Jérôme (1992), 'Emerging high-tech firms: how durable are their comparative start-up advantages?', *Journal of Business Venturing*, **7**(4), 303–22
Doutriaux, Jérôme (1998), 'Canadian science parks, universities, and regional development', in John de la Mothe and Gilles Paquet (eds), *Local and Regional Systems of Innovation*, Boston, Mass.: Kluwer Academic Publishers, pp. 303–24.
Doutriaux, Jérôme (1999), 'Technology incubation in Canadian science parks: two case studies', in L. Corona (coord.), 'Experiencias, retos y oportunidades regionales en la innovación tecnológica', Universidad Autonoma del Estado de Hidalgo, Pachuca, Mexico, 14-15 November, 357–69.
Doutriaux, Jérôme, (2000), 'Private sector financed research activities at universities', in J.A. Holbrook and D.A. Wolfe (eds), *Innovation, Institutions, and Territory, Regional Innovation Systems in Canada*, Montreal: McGill-Queen's University Press, pp. 93–123.
Doutriaux, Jérôme (2003), 'University–Industry Linkages and the Development of Knowledge Clusters in Canada', *Local Economy*, **18**(1), 63–79.
Economist Intelligence Unit (2005), '*The World in 2005*', *The Economist*, UK.
Economist (The) (2004), 'The peso crisis, ten years on Tequila slammer', 29 December, Mexico City. From *The Economist print edition*.
Edquist, C. (ed.) (1997), *System of Innovation, Technologies, Institutions, and Organizations*, London: Pinter.
Etzkowitz H. and Loet Leydesdorff (1998), 'A triple helix of university–industry–government relations', *Industry & Higher Education*, August, **12**(4), 197–201.
Europa World Yearbook, Volume II The (2003), 'The United States of America: introductory survey', London and New York: Europa Publications, pp. 4401–19.
FP Markets, Canadian Demographics (2001), Toronto, Ontario: Financial Post Publications.
Freeman, Chris (1999), 'Innovation systems: city-state, national, continental and sub-national', in J.E. Cassiolato and Helena Lastres (eds), *Globalizaçao & Inovaçao Localizada. Experiências de Sistemas Locais no Mercosul*, Brazil: IBICT-MCT Brasília, Brazil.
Gacel-Ávila, Jocelyne (2001), *La Internacionalización de las Universidades Mexicanas. Políticas y Estrategias Institucionales*, ANUIES.
Globe and Mail, The (2000a), Toronto, Ontario, 24 November.

Hansen, M.T., H.W. Chesbrough, N. Nohria and D.N. Sull (2000), 'Networked Incubators, Hothouses of the New Economy', *Harvard Business Review*, September–October.
Harms, D., R. Girard and B. Peterman (2001), 'Creating Economic Activity: University of Saskatchewan example', *UST*, October.
Industry Canada (2001), 'The ICT Sector in Canada', August, Industry Canada, http://strategis.ic.gc.ca/pics/it/sp1199e.pdf.
Industry Canada (2004), 'ICT Sector Intramural R&D Expenditures, 2004 Intentions', http://strategis.ic.gc.ca/epic/internet/inict-tic.nsf/en/h_it05385e.html, consulted on 25 February, 2005.
INEGI (2001), *XII Censo General de Población y Vivienda 2000*, México.
IRAP (2000), 'Performance report 1999–2000: building capacity in the Canadian system of innovation', August 2000, p. 19, http://irap-pari.nrc-cnrc.gc.ca/pr99—c.pdf.
Keeble, David and Frank Wilkinson (2000), 'High-technology SMEs, regional clustering and collective learning: an overview', in David Keeble and Frank Wilkinson (eds), *High-Technology Clusters, Networking and Collective Learning in Europe*, Burlington, US: Ashgate, pp. 1–20.
Klein, Hans (2001), 'Technology push-over: defense downturns and civilian technology policy', *Research Policy*, **30**, 937–51.
Kogut, B. (ed.) (1991), 'Country capabilities and permeability of borders', *Strategic Management Journal*, **12**, 33–47.
Krugman, Paul (1991), 'Increasing returns in economic geography', *Journal of Political Economy*, **99**, 483–99.
Landry, Julie (2000), 'Econets suffer from fatal flaw', *Redherring*, 27 November.
Langford, Cooper H., J.R. Wood and T. Ross (2002), 'Origins and structure of the Calgary Wireless Cluster', University of Calgary, *Faculty of Communication and Culture* September, http://www.thecis.ca/working_papers1.html
Larson, Charles F. (2001), 'U.S. industry moderates its R&D spending', *Research-Technology Management*, **44**(4), July–August, pp. 2–4.
Laurin, Annie (2001), 'Pleins feux sur Québec, une capitale à vocation technologiquie', *Le Soleil*, Quebec, Canada, 7 October.
Lavrow, M. and S. Sample (2000), 'Business incubation, trend or fad?', University of Ottawa, E-MBA program report, October.
Lederman, Daniel, William F. Maloney and Luis Servén (2005a), 'Lessons from NAFTA for Latin America and the Caribbean Countries', chapter 1 in *Lessons from NAFTA for Latin America and the Caribbean*, The World Bank, pp. 1–26.
Lederman, Daniel, William F. Maloney and Luis Servén (2005b), 'Innovation in Mexico: NAFTA is not enough', chapter 6 in *Lessons from NAFTA for*

Latin America and the Caribbean, The World Bank, pp. 247–88.
Leydesdorff, L. and Henry Etzkowitz (1998), 'The triple helix as a model for innovation studies', *Science and Public Policy*, June, **25**(3), 195–203.
Lugar, Michael and Harvey Goldstein (1991), *Technology in the Garden: Research Parks and Regional Economic Development*, Chapel Hill, NC: The University of North Carolina Press.
Lukas, R. (1988), 'On the mechanics of economic development', *Journal of Monetary Economics*, **22**(1), 3–42.
Marshall, Alfred ([1890], 1961), *Principles of Economics*, London: Macmillan.
Martínez Rizo, Felipe (2000), *Nueve Retos para la Educación Superior. Funciones, Actores y Estructuras*, ANUIES, Mexico.
Master, J.W. and C. Reichert (2000), 'Calgary, Heart of the New West', presentation by J.W. Master, president and CEO of Calgary Technologies Inc., and by Charles Reichert, Director Infoport, July.
Mian, Sarfraz (1994), 'U.S. university sponsored technology incubators: an overview of management, policies and performance', *Technovation*, Elsevier Advanced Technology, **14**(8), 515–28.
Mian, Sarfraz (1996), 'Assessing value-added contributions of university technology business incubators to tenant firms', *Research Policy*, **25**, 325–35.
Mian, Sarfraz (1997), 'Assessing and managing the university technology business incubator: an integrative framework', *Journal of Business Venturing*, **12**(4), 251–340.
Mian, Sarfraz and Walter Plosila (1996), 'Emerging models of state programs in technology and enterprise development: the US experience', *INFORMS annual conference*, Atlanta.
Mian, Sarfraz and Walter Plosila (1997), 'Mechanisms for commercializing university research: a study of selected university programs in the U.S.', paper presented at *The Sixth International Conference on Management of Technology*, Gothenburg, Sweden.
Minnesota Office of Science and Technology (1988), *State Technology Programs in the United States – 1988*. Prepared by Minnesota Department of Trade and Economic Development in Cooperation with LFW Management Associates, Alexandria, VA.
Moulaeert, Frank and Faridah Djellal (1995), 'Information technology consultancy firms: economies of agglomeration from a wide-area perspective', *Urban Studies*, **323**(1), 105–122.
Mowery, David and Nathan Rosenberg (1993) 'The U.S. National Innovation System', in Richard Nelson (ed.), *National Innovation Systems*, New York: Oxford University Press.
Musalem, Omar (1989), *Innovación tecnológica y parques científicos.*

Ensayos sobre ciencia y tecnología, México: Nacional Financiera.

National Academy of Engineering (1988), *The Technological Dimensions of International Competitiveness*, Washington, DC.

National Governors' Association (2002), *New York Profile of the State Economy*.

National Science Board (2000), *National Science and Engineering Indicators 2000*, National Science Foundation, Washington, DC: Government Printing Office.

National Science Board (2004), *Science and Engineering Indicators 2004*, National Science Foundation, Arlington, VA, USA.

National Science Foundation (1998), *Venture Capital Investment Trends in the United States and Europe*, NSF 99-303.

National Science Foundation (2000), *Science and Technology Pocket Book*, NSF.

National Science Foundation (2001), *Science and Engineering Indicators 2000*, NSF.

National Science Foundation (2002), *State Science and Technology Indicators*, National Science Foundation.

National Science Foundation (2004), *Science and Engineering Indicators*, Washington DC: Government Printing Office.

National Science Foundation (NSF) and National Science Board (NSB) (1996), *Science and Engineering Indicators*, Washington DC: Government Printing Office.

National Venture Capital Association (2005), website accessed February 1, http://www.nvca.org/nvca7_29_03.html.

Nelson, Richard (ed.) (1993), *National Innovation Systems: A Comparative Analysis*, New York: Oxford University Press.

NY State Science & Technology Foundation (1992), *New York State Centers for Advanced Technology Program: Evaluation, Past Performance and Preparing for the Future*, Menlo Park: CASRI International.

OECD (1997a), *The Measurement of Scientific and Technological Innovation Data. Proposed Guidelines for Collecting and Interpreting Technological Innovation Data: OSLO MANUAL*, European Commission /Eurostat.

OECD (1997b), *Technology Incubators, Nurturing Small Firms*, Paris: OECD.

OECD (1997c), *National Innovation Systems*, Paris: OECD.

OECD (1999), *Business Incubation. International Case Studies*, Paris: OECD.

OECD (2001a), *Innovative Clusters: Drivers of National Innovation Systems*, Paris 15 July, OECD.

OECD (2001b), *Innovative Network: Cooperation in National Innovation Systems*, Paris: OECD.

OECD (2001c), *Innovative People. Mobility of Skilled Personnel in National Innovation Systems. Science and Innovation*, Paris: OECD.

OECD (2002a), *Benchmarking Industry–Science Relationships*, Paris: OECD.
OECD (2002b), *Dynamising National Innovation Systems*, Paris: OECD.
OECD (2003a), *Turning Science into Business: Patenting and Licensing at Public Research Organisation*, Paris: OECD.
OECD (2003b), *OECD Science, Technology and Industry Scoreboard 2003*, Paris: OECD.
OECD (2003d), *Economic Survey of Mexico*, Policy Brief November. Paris: OECD.
OECD (2004a), *Science and Innovation Policy. Key Challenges and Opportunities*, Meeting of the OECD Committee for Scientific and Technology, Policy at Ministerial Level, 29–30 January, Paris: OECD.
OECD (2004b), *Science and Technology Statistical Compendium*, Meeting of the OECD Committee for Scientific and Technology, Policy at Ministerial Level, 29–30 January, Paris: OECD.
OECD (2004c), *Main Science and Technology Indicators 2004/2*, Paris: OECD.
OECD (2004d), *OECD Science, Technology and Industry Outlook*, Paris: OECD.
OECD (2004e), *Territorial Reviews: Mexico City*, October, Policy Brief, Paris: OECD.
OECD and CERI (2004), *National Review on Educational R&D Examiners*, Report on Mexico, Paris: OECD.
Ottawa Business Journal (2002), 18 February.
Ottawa Economic Development Corporation, '1998 Ottawa facts', Ottawa, Canada.
Partida Romo, Raquel (1999), 'Impacto e Innovación tecnológica de la industria electrónica en Guadalajara, en la década de los noventa', in Leonel Corona (coord.), *Innovación tecnológica y desarrollo regional*, Universidad Antónoma del Estado de Hidalgo, Pachuca, Mexico, pp. 189–208.
Pavitt, K. (1984), 'Sectorial pattern of technical change: toward a taxonomy and a theory', *Research Policy*, **13**, North Holland, 343–73.
Pecyt (2001), *Programa Especial de Ciencia y tecnología 2001–2006,* SEP-Conacyt, Mexico, 30 October.
Plosila, Walter (1988), 'Technological innovation: the sub-national government experience in the United States', paper presented at *Technology and the City: An International Conference on the Potential of New Technology Enterprises to Regenerate Inner Cities*, London.
Plosila, Walter (2004), 'State science and technology-based economic development policy: history, trends and developments, and future directions', *Economic Development Quarterly*, **18**(2), 113–26.
Ponce Ramirez, Luis (1999), 'El sistema regional de ciencia y tecnología', in Corona Leonel (coord.), *Innovación Tecnológica y Desarrollo Regional*,

Pachuca, México: UAEH.
Porter, Michael (2001), 'Strategy and the internet', *Harvard Business Review*, March, pp. 63–80.
Porter, Michael and Scott Stern (2002), 'National Innovation Capacity', in Porter et al. (eds), *The Global Competitiveness Report 2001–2002*, New York: Oxford University Press, pp. 102–119.
Reamer, Andrew, Larry Icerman and Jan Youtie (2003), 'Technology transfer and commercialization: their role in economic development', US Department of Commerce, EDA report number 990607435.
Reséndiz Nuñez, Daniel (2000), *Futuros de la Educación en México*, Mexico: Siglo XXI.
Rice, M. (1993), 'Intervention mechanisms used to influence the critical success of new ventures: an exploratory study', Unpublished Ph.D. Dissertation, Renssalaer Polytechnic Institute, Troy, NY.
Rifkin, Jeremy (2000), *The Age of Access*, New York: Jeremy P. Tarcher/ Putman Inc.
Roberts, Edward and Denis Malone (1996), 'Policies and structures for spinning-off new companies from research and development organizations', *R&D Management*, **26**(1), 17–48.
Rodriguez, J. and M. Villa (1997), 'Dinámica sociodemográfia de las metrópolis latinoamericanas durante la segunda mitad del siglio XX', Natas de Problación, **65**, Junio, Santiago de Chile: CEPAL.
Romer, Paul (1986), 'Increasing returns and long-run growth', *Journal of Political Economy*, **94**, October, 1002–37.
Rosenberg, N. (1972), *Technology and American Economic Growth*, New York: Harper and Row.
Rothwell, Roy (1994), 'Towards the fifth generation innovation process', *International Marketing Review*, **1**(1), 7–31.
Ryan, Vincent (2000), 'After the incubator', *Telephony*, 20 November, 2000.
Sanz, Luis (2004), 'The future role of science parks in metropolitan science regions', SSES and ESBRI Research Seminar. Stockholm, March. http://www.iasp.ws/information/definitions.
Sanz, Luis and Philip Cooke (2004) 'Regional innovation systems: an evolutionary approach', in Cooke et al. (eds), *Regional Innovation Systems: The Role of Governance in a Globalized World*, 2nd edn, London, New York: Routledge Press.
Saxenian, Annalee (1994), *Regional Advantage: Culture and Competition in Silicon Valley and Route 128*, Cambridge, Mass.: Harvard University Press.
Schoonhoven, C.B. and E. Romanelli (eds) (2002), *The Entrepreneurship Dynamics: Origins of Entrepreneurship and the Evolution of Industries*, Stanford, CA: Stanford University Press.
Shapira, Philip (2001), 'US manufacturing extension partnerships: technology

policy reinvented', *Research Policy*, **30**, 977–92.

Sherman, Hugh and David Chappell (1998), 'Methodological challenges in evaluating business incubator outcomes', *Economic Development Quarterly*, **12**(4) November, 313–21.

Smilor, R. and M. Gill (1986), *The New Business Incubator: Linking Talent, Technology and Know-How*, Mass.: Lexington Books.

SSTI Weekly Digest (1997), 'Memorandum of Understanding: Officially Established', *USIP*, 27 June, State Science and Technology Institute, Ohio.

SSTI Weekly Digest (1998), 'State science and technology institute', Columbus, Ohio, USA, February 6, http://www.ssti.org.

Statistics Canada, *Service Bulletin on Science Statistics*, Science, Innovation and Electronic Information Division, 88-001-XIE, Ottawa, Canada, various issues.

Statistics Canada, Provincial Gross Domestic Product by Industry, Catalog 15-203-XPB, Ottawa, Canada, various issues.

Statistics Canada (1999), *Education Indicators in Canada*, catalog 81-582-XIE.

Statistics Canada, CANSIM, Ottawa, Canada.

Storper, Michael (1997), 'The regional world, territorial development in a global economy', New York, London: Guildford Press.

Stuart, T. and O. Sorenson (2002), 'The geography of opportunity: spatial heterogeneity in founding rates and the performance of biotechnology firms', *Research Policy*, **32**, 229–53.

Stuart, T. and O. Sorenson (2003), 'The geography of opportunity: spatial heterogeneity in founding rates and the performance of biotechnology firms', *Research Policy*, **32**, 229–53.

Tonatzky, L., Y. Batts, N. McCrea, M. Lewis and L. Quittman (1996), *The Art and Craft of Technology Business Incubation*, Southern Technology Council, Durham, NC and NBIA, Athens, OH.

Turner, Barry (ed.) (2002), *The Statesman's Yearbook: the Politics, Cultures and Economics of the World*, London and New York: Palgrave, p. 1731.

Unger, Kurt and Mateo Oloriz (1998), *Innovation and Foreign Technology in Mexico's Industrial Development*, Mexico, CIDE Num. 117.

Unger, Kurt and Mateo Oloriz (2000), 'Globalization of production and technology', in M. Cimoli, *Developing Innovation Systems. Mexico in a Global Context*, London and New York: Continuum.

US Census Bureau (2004), 'State and country quickfacts', visited 1 February, 2005, http://quickfacts.census.gov/qfd/states/00000.html.

U.S. Department of Commerce (1997), *The Global Context for US Technology Policy*, Washington, DC: Office of Science and Technology Policy.

US Dept of Commerce (1997–98), *State and Metropolitan Area Data Book*, Washington, DC.

Yin, Robert (1984), *Case Study Approach*, Beverley Hills, CA: Sage.
Zambrano Plant, Carlos (2001), 'La promoción económica de una región de alto crecimiento industrial: el caso de Nuevo León', *El Mercado de Valores*, NAFIN, Mexico, November, pp. 31–8.

WEB REFERENCES

http://cde.itesm.mx/red_directorio.php
http://content.calgary.ca/CCA/City+Hall/Business+Units/Customer+Service+and+Communications/Corporate+Marketing/Municipal+Handbook/+Welcome+to+Calgary/Industrial+Base.htm#research percent20and percent20 development
http://irap-pari.nrc-cnrc.gc.ca
http://precarn.ca/Corporate/corporate.cfm
http://strategis.ic.gc.ca/SSG/bo01544e.html, Overview of Biotechnology, 8 August, 2001.
http://strategis.ic.gc.ca/epic/internet/inict-tic.nsf/en/h_it05385e.html
http://strategis.ic.gc.ca/pics/it/it05373e.pdf; http://strategis.ic.gc.ca/pics/it/sp1199e.pdf, The ICT Sector in Canada, August 2001.
http://194.30.15.20/iaspworld/about/fabout.htm, International Association of Science Parks.

Next addresses are http://

www.accc.ca/english/Colleges/membership_list.cfm
www.agwest.sk.ca/agwest.shtml
www.areadevelopment.com/Pages/Features/Feature11.html
www.aucc.ca/en/aboutindex.html
www.bancomext-mtl.com/invest/vox128.htm
www.calgarytechnologies.com.
www.canarie.ca/about/about.html.
www.cfyge-prifepei.ipn.mx
www.cihr.ca/
www.city.saskatoon.sk.ca, quick facts, June 2002.
www.conacyt.mx/dat/avance);
www.conacyt.mx/dadrys/directorio/index.html
www.cqib.org
www.cvca.ca Venture Capital Association of Canada
www.entrepreneurship.com
wwww.gov.sask.ca/bureau.stats, Jan 2005
www.gov.sk.ca/econdev/investment/sixsctrs/technology/sector.shtm

www.gov.sk.ca/soco/faq.htm
www.inegi.gob.mx
www.infostat.gouv.qc.ca
www.innovationplace.com/html/frameset.html
www.mic.gouv.qc.ca/PME-REG/regions/, Quebec Government data.
www.nce.gc.ca/en/netseng.ht
www.nrc-cnrc.gc.ca
www.ocri.ca
www.ocri.ca/ocrimodel/publications/1204_ottawafacts.pdf
www.oecd.org/department/0,2688,en_2649_34357_1_1_1_1_1,00.html
www.parctechno.qc.ca/english/histo.htm
www.researchinfosource.com/top50.html
www.siicyt.gob.mx
www.soco.sk.ca/links/index.html
www.speqm.qc.ca/speqm-fra/cont-quebec-technoregion.html
www.speqm.qc.ca/speqm-fra/cont-quebec-technoregion.html
www.sreda.com/science-city_biotechnology.php
www.statcan.ca. Statistics Canada.
www.statcan.ca/english/census96/apr14/educ.htm.
www.statcan.ca/english/Pgdb/econ05.htm
www.statcan.ca/english/Pgdb/gblec02a.htm
www.techba.com
www.thecis.ca/working%20papers/Origins%20and%20Structure.pdf
www.wto.org, International Trade Statistics 2001.

Index